CLIMATE CHANGE MITIGATION AND INTERNATIONAL DEVELOPMENT COOPERATION

Asia's share of the world's carbon dioxide emissions from fuel grew from 23 percent in 1990 to 28 percent in 2008. Asian countries are in the process of reducing their amount of greenhouse gas emissions per unit of GDP. The problem is that these efforts alone would be inadequate if they were conducted alongside business as usual. Whether developing countries can leapfrog from a business-as-usual situation to a low carbon society depends on whether or not they can procure the required funds and technology. Through development policy lending and the like, with a focus on climate change policies, if it is possible to actively incorporate the low carbon perspective of NAMAs into development strategies, funding will contribute to energy conservation and new energy market expansion in the countries concerned.

Climate Change Mitigation and International Development Cooperation has two parts. Part I reviews the current status of, and challenges for creating, an international framework on mitigation. Part II introduces examples from developing countries and ends by introducing initiatives to create a roadmap for moving towards a low carbon society in Japan.

This book is a summary of knowledge about climate change mitigation for practitioners involved in countermeasures in developing country governments, development assistance organizations and non-governmental organizations.

Ryo Fujikura is Professor at the Faculty of Humanity and Environment, Hosei University, and Visiting Fellow, JICA Research Institute.

Tomoyo Toyota is a Research Associate at JICA Research Institute.

CLIMATE CHANGE MITIGATION AND INTERNATIONAL DEVELOPMENT COOPERATION

Edited by Ryo Fujikura and Tomoyo Toyota

Routledge
Taylor & Francis Group
LONDON AND NEW YORK

JICA Research Institute

First published 2012
by Routledge
2 Park Square, Milton Park, Abingdon, Oxon OX14 4RN

Simultaneously published in the USA and Canada
by Routledge
711 Third Avenue, New York, NY 10017

Routledge is an imprint of the Taylor & Francis Group, an informa business

British Library Cataloguing in Publication Data
A catalogue record for this book is available from the British Library

Library of Congress Cataloging in Publication Data
Library of Congress Cataloging-in-Publication Data
Toyota, Tomoyo, 1981–
Climate change mitigation and international development cooperation / Tomoyo Toyota, Ryo Fujikura.
p. cm.
Includes bibliographical references and index.
1. Greenhouse gas mitigation–Asia. 2. Greenhouse gas mitigation–Developing countries. 3. Greenhouse gas mitigation–International cooperation. 4. Climate change mitigation–Asia. 5. Climate change mitigation–Developing countries. 6. Climate change mitigation–International cooperation. I. Fujikura, Ryo, 1955– II. Title.
QC981.8.C5T696 2012
363.738'746095–dc23

2011041904

ISBN: 978-0-415-50864-3 (hbk)
ISBN: 978-0-415-50865-0 (pbk)
ISBN: 978-0-203-12366-9 (ebk)

Typeset in Bembo
by Wearset Ltd, Boldon, Tyne and Wear

CONTENTS

List of illustrations vii
List of contributors xi
Preface xiii

1 Introduction: Challenges for Japan, Challenges for Asia 1
Ryo Fujikura

PART I
International framework on mitigation **17**

2 A Review of Discussions on an International Framework to Address Climate Change, and Mitigation Actions by Developing Countries 19
Yasuko Kameyama

3 Current Status and Critical Issues of the CDM 37
Kazuhito Yamada and Mariko Fujimori

4 REDD+: A New Strategy for Conserving Tropical Forests: Scientific Background of REDD+ and International Discussions 49
Mitsuo Matsumoto

5 International Intercity Collaboration and its Potential Application in Low Carbon Development 61
Hidenori Nakamura and Hideyuki Mori

PART II
Cases of climate change mitigation 97

6 Sectoral Policies for GHG Mitigation in Asia: Policies for Energy, Industry, Transportation, and Buildings in China and India 99
Frank Hiroshi Ling

7 Development of a Nationally Appropriate Mitigation Action Plan (NAMA): A Case Study of Thailand's Waste and Wastewater Management Sector 120
Kazuhito Yamada and Mariko Fujimori

8 Effectiveness of Development Policy Lending in Mainstreaming and Financing Climate Change: Case Studies from Indonesia and Vietnam 149
Taro Katsurai and Yuka Murakami

9 Carbon Dioxide Reduction Effects of Yen Loan Coal-fired Power Plant Projects 180
Tomoyo Toyota

10 Rural Electrification in Bangladesh: Implications for Climate Change Mitigation 202
Shinji Kaneko, Satoru Komatsu, and Partha Pratim Ghosh

11 Asian Low-Carbon Society Scenario Development and Policy Implementation 227
Junichi Fujino

Commentary to Parts I and II: "Not Such Strange Bedfellows: Climate Mitigation as a Buttress for Sustainable Development" 241
Sandra S. Nichols, John A. Pendergrass, and J. Cory Connolly

Index 255

ILLUSTRATIONS

Figures

2.1	CO_2 emissions from fuel combustion by countries in 2007	28
2.2	Year 2004 distribution of regional per capita GHG emissions over the population of different country groupings	29
2.3	Correlation between climate change and poverty	33
3.1	Outline of the CDM	38
3.2	Concepts of emission reduction	41
3.3	Relationship between estimated CERs and GDP	46
4.1	Concept of emission reductions in REDD+	50
4.2	Global carbon cycle for the 1990s	50
4.3	Cumulative mitigation potential according to mitigation options under the 2.7 US$/t$CO_2$ +5%/yr annual carbon price increment	53
4.4	Basis of estimation methods of forest carbon stocks for REDD+	55
4.5	Estimation methods of carbon stock changes in forest biomass	55
4.6	Comparison of methods of setting reference levels and emission reductions to be assessed	57
5.1	Distribution of the frequency of cities' participation in network events by network programme	68
5.2	Concept of a platform for international intercity low carbon collaboration in Asia	91
6.1	Share of incremental energy demand, imports, and energy related CO_2 emissions growth between 2000 and 2006	100
6.2	Growth of China's installed wind capacity	104
7.1	Waste management sector activities	128

8.1 Indonesian emission growth projection from 2005 to 2030 154
8.2 Structure of CCPL policy matrix (2007–2009) 156
8.3 Implementation structure of CCPL 160
8.4 Structure of CCPL policy matrix 162
8.5 Management and implementation arrangement of SP-RCC 171
9.1 GHG emissions by sector in OECD and non-OECD countries 184
9.2 Power generation structure change 188
9.3 CO_2 emissions per KWh of electricity and heat output 189
9.4 Power generation change according to fuel and share of power generation structure 191
9.5 Thermal efficiency according to fuel 192
9.6 National averages of coal-fired power plant thermal efficiencies 192
10.1 Share of energy related CO_2 emissions in electricity and heat sector to total emissions 202
10.2 PBS organizational chart 206
10.3 Comparison of performance indicators 208
10.4 Comparison of monthly payments on SHS and energy savings 217
10.5 Incomes of non-SHS households and SHS households before installation 220
10.6 Ratio of energy related CO_2 emissions in electricity and heat sector to total emissions for Bangladesh, India, and China 221
10.7 International cooperation and BOP in learning investment for new PV technology 223
11.1 Multiple perspectives are necessary to consider mid- and long-term mitigation roadmaps 231
11.2 Contribution by mitigation measures in the Shiga scenario 233
11.3 Six actions in environmental modeling City Kyoto plan 235
11.4 Low carbon cities policy packages in Iskandar Malaysia 237
11.5 Workshop on low carbon Asia research project 238

Tables

2.1 History of climate change negotiations 20
2.2 Emission targets of countries in Asia and the Pacific region 31
3.1 Types, scales, and activities of the CDM projects 39
3.2 Share of registered CDM projects categorized by host country 43
3.3 Share of registered CDM projects categorized by project type 44
3.4 Share of CERs of registered CDM projects categorized by project type 44

3.5	Estimated CERs per project types	44
4.1	Annual change in forest area by region and subregion, 1990–2010	51
4.2	Ten countries with largest annual net loss of forest area, 1990–2010	52
4.3	Results of a preliminary estimation of forest carbon stocks in Cambodia	56
5.1	Breakdown of activities of four international environmental networks	65
5.2	Kitakyushu Initiative activities	65
5.3	CAI-Asia activities	66
5.4	ICLEI activities	66
5.5	CITYNET activities	67
5.6	Frequency of cities' participation in network events	69
5.7	Top five cities in terms of frequency of participation in network events	69
5.8	State of intercity mutual reference (CITYNET)	71
5.9	State of intercity mutual reference (ICLEI Southeast Asia)	71
5.10	State of intercity mutual reference (Kitakyushu Initiative)	72
5.11	Formulation of knowledge networks for sustainable development	76
5.12	State of Japanese local governments' engagement in the JICA technical cooperation programme for the environment (2007–2009)	82
5.13	Low carbon development policies and practices of Japanese municipal government as of 2008	83–4
5.14	Low carbon development policies and practices of local governments in East Asian developing countries	84
5.15	Preliminary assessment of intercity collaboration on low carbon development in Asia	92
6.1	Savings in primary energy consumption and emissions in China between 2005 and 2008	102
7.1	Examples of NAMAs submitted by non-Annex 1 Parties	123–5
7.2	Step 1: reviewing related policies/plans and their progress in NAMA sector	129–31
7.3	Step 2: setting reference scenarios in the NAMA sector	132–7
7.4	Step 3: setting NAMA targets in the NAMA sector	139–41
7.5	Selection methodology for priority NAMA projects (example)	142
7.6	Steps 4, 5, 6, and 7	143–4
7.7	Summary of selected countermeasures	146–7
8.1	Summary of GHG emissions from 2000–2005 for all sectors	153
8.2	Baseline and Main Policy Reform (2007–2009)	158–9

8.3	Main policy actions (2010)	162
8.4	GHG emissions of Vietnam, 1994 and 2000	166–7
8.5	Main policy reforms for the first and second year	170
9.1	Yen loan projects examined in this study	183
9.2	Basic data on energy, economy, and emissions by region, 1970–2007	186–7
9.3	Electrification rates and population without electricity in 2008	188
9.4	Thermal efficiency of fuels	190
9.5	Thermal efficiency in reference scenarios	196
9.6	Estimated GHG emission reduction effects	197–8
9.7	Cumulative CO_2 emissions in Japan and cumulative CO_2 reduction effects from 2000 to 2007	199
9.8	Yen loan projects in the power generation sector	200
10.1	Electricity capacity and generation in 2008 and 2035 in OECD and non-OECD countries	203
10.2	Installation of SHS by year	211
10.3	Contract packages of SHS	212
10.4	Common reasons and number of respondents	212
10.5	Number of mobile phones	213
10.6	Distance and the frequency of visits to place for recharging mobile phones	213
10.7	Determinants of decision for SHS adoption	214
10.8	Determinants of decision for SHS size selection	215
10.9	Comparison of monthly expenditure on SHS and fuel cost reduction	217
10.10	Comparison of donor priorities in electrification projects	218
10.11	Record of support for IDCOL by each donor agency	219
11.1	Stakeholders' actions to develop sustainable society scenarios and roadmaps	232

CONTRIBUTORS

J. Cory Connolly (Mr.) Research Associate, Environmental Law Institute (USA).

Ryo Fujikura (Mr.) Professor, Faculty of Humanity and Environment, Hosei University, and Visiting Fellow, JICA Research Institute (Japan).

Mariko Fujimori (Ms.) Director, PC-Institute for Global Environment Research, Pacific Consultants Co., Ltd. (Japan).

Junichi Fujino (Mr.) Senior Researcher, National Institute for Environmental Studies (Japan).

Partha Pratim Ghosh (Mr.) Information Management Specialist, Global Communication Center (Bangladesh).

Akio Hosono (Mr.) Director, JICA Research Institute (Japan).

Yasuko Kameyama (Ms.) Senior Researcher, National Institute for Environmental Studies (Japan).

Shinji Kaneko (Mr.) Professor, Hiroshima University (Japan).

Taro Katsurai (Mr.) Senior Expert, International Development Associates, Ltd. (Japan).

Satoru Komatsu (Mr.) Assistant Professor, Hiroshima University (Japan).

Frank Hiroshi Ling (Mr.) Associate Researcher, Institute for Global Environmental Strategies (Japan), and Researcher, Institute for Global Change Adaptation Science, Ibaraki University (Japan).

Mitsuo Matsumoto (Mr.) Principal Research Coordinator and Director of REDD Research and Development Center, Forestry and Forest Products Research Institute (Japan).

Hideyuki Mori (Mr.) President, Institute for Global Environmental Strategies (Japan).

Yuka Murakami (Ms.) Senior Project Formulation Advisor, Japan International Cooperation Agency (Japan).

Hidenori Nakamura (Mr.) Associate Researcher, Institute for Global Environmental Strategies (Japan).

Sandra S. Nichols (Ms.) Senior Attorney, Environmental Law Institute (USA).

John A. Pendergrass (Mr.) Senior Attorney, Environmental Law Institute (USA).

Tomoyo Toyota (Ms.) Research Associate, JICA Research Institute (Japan).

Kazuhito Yamada (Mr.) Deputy Director, New Business Development Division, Pacific Consultants Co., Ltd. (Japan)

PREFACE

It is with great pleasure that we bring to publication the product of our research on *Climate Change Mitigation and International Development Cooperation.* This is the second publication of work on climate change by the Japan International Cooperation Agency Research Institute (JICA-RI), following the 2010 publication entitled *Climate Change Adaptation and International Development: Making Development Cooperation More Effective*, which was also published by Earthscan.

Coping with climate change through mitigation and adaptation is undoubtedly one of the world's most urgent challenges. It is essential that we make every possible effort to reduce emission of greenhouse gases into the atmosphere. An equally important task is to devise and implement measures for coping with the actual and anticipated impact of the phenomenon.

In this context, please allow me to share with the reader what I believe are valuable lessons that Japan has to offer concerning energy conservation and climate change mitigation.

In Japan, public awareness of environmental issues heightened gradually during the 1960s as air and water pollution worsened with accelerated industrialization. The subsequent 1973 oil shock was a major turning point, in that it forced Japanese companies and the general public to take measures for improving energy efficiency. Energy conservation has constantly been an issue in Japan ever since, and today Japan is one of the most energy efficient country in the world.

For example, Japan's energy efficiency coefficient (energy consumption divided by gross domestic product (GDP)) is the lowest among major industrial countries. Japan's figure is half that of the United States and one-seventh that of China. According to *General Energy Statistics* published by Japan's Agency for Natural Resources and Energy, energy efficiency in Japan improved 37 percent between 1973 and 2003, going from 1.6 to 1.07 kilotons in petroleum-equivalent per JPY 1 billion of GDP annually.

Efforts by the manufacturing sector have been impressive. Energy intensive industries such as steel, chemical products, and paper have achieved major improvements. Other industries have used both technological innovation and incremental improvements to enhance energy efficiency. As a part of this process, Japan introduced a technique called the "top-runner program," the idea of which was to set the energy standard for a variety of appliances based on the most energy efficient product available on the market. This approach was later praised by Lester Brown as "the most effective energy saving initiative ever introduced in the world" (*Plan B 4.0: Mobilizing to Save Civilization*, 2009). For example, energy efficiency of gasoline-powered passenger vehicles improved by 22.8 percent between 1995 and 2005. Energy savings in air conditioners and electric refrigerators was even more dramatic, improving by 67.8 percent and 55.2 percent, respectively, during the period from 1997 to 2004. Thus, 21 major energy consuming products of that period were targeted using the leading-runner approach.

As a result of efforts like these, the potential for further reduction of CO_2 emissions in Japan had come to be seen as much more limited than other countries, as some believed that earlier measures for energy conservation had achieved all that could be done. The events of March 11, 2011, however, and the subsequent Fukushima Daiichi nuclear disaster changed all that. Faced with acute shortage of electricity and to avoid unplanned blackouts, numerous electricity conservation measures were introduced, leading both the general public and private enterprise to review their lifestyles and modes of operations and conserve electricity as much as possible. Lo and behold, at the time of this writing, the government, private enterprise, and the general public together have attained a 15 percent reduction of electric power usage in areas serviced by Tokyo Electric Power Company during the months of July and August of 2011.

Looking back on the Japanese experience over the past five decades and particularly during the arduous period following the 3/11 natural disasters, one can conclude that, given effort and ingenuity, there are infinite possibilities for energy conservation and, hence, for climate change mitigation. This, I believe, is an encouraging message that can and should be borne in mind by all of us when discussing climate change mitigation.

Although Japan now faces what former Prime Minister Naoto Kan has described as "the toughest and the most difficult crisis ... in the 65 years after the end of World War II," Mr. Kan has also noted that Japan remains committed to continuing its "contribution to the development of the international community" and tackling climate change. I am glad that my institute can join in that important endeavor with the publication of this book, and hope that it will help inform debate on climate change mitigation policies.

Akio Hosono
Director
JICA Research Institute

1

INTRODUCTION

Challenges for Japan, challenges for Asia

Ryo Fujikura

I would like to extend my deepest condolences to the victims of the Great East Japan earthquake, which struck in March 2011. As a Japanese citizen, I also wish to express my sincere appreciation to all the individuals, organizations, corporations, and governments from 163 countries and regions, as well as 43 international organizations, that have provided support, both physical and emotional, to the victims of this disaster.

Challenges for Japan

March 11, 2011 is a day the Japanese people will never be able to forget. On that day, the eastern shores of Japan's Tohoku (northeast) region were struck by a tsunami up to ten meters high, generated by what was reportedly a magnitude 9.0 earthquake – among the largest possible – that occurred below the seafloor about 130 kilometers offshore. More than 20,000 people died or went missing. Although there were fewer victims than the Sumatra earthquake and tsunami of 2004 (when 220,000 died), and the Haiti earthquake in 2010 (230,000 died), the Great East Japan earthquake has had an additional and unexpected impact on the world: the earthquake and subsequent nuclear accident exposed for all to see that the safety of nuclear power is a myth.

Before the accident, the Tokyo Electric Power Company (TEPCO), which supplies the greater Tokyo area, had the capacity to supply 57,000 megawatts of electricity. Of this total, three nuclear power plants (Fukushima Daiichi, Fukushima Daini, and Kashiwazaki) provided 17,300 megawatts. The earthquake and tsunami caused a meltdown at three of the six reactors at Fukushima Daiichi nuclear power plant. No major damage occurred at Fukushima Daini, although operations were completely halted. Due to damage that also occurred at a number of thermal power plants, the electricity supplied by TEPCO dropped to 31,000 megawatts immediately after the disaster (Ministry of Economy, Trade and Industry 2011). TEPCO was able to avoid the crisis of a massive blackout in the greater Tokyo area thanks to the emergency release of water that had been stored for hydropower production, but still implemented rolling power blackouts starting on March 13. This was the first time scheduled power blackouts had occurred since just after the end of World War II.

TEPCO was able to restore its generating capacity and stop the rolling power blackouts in April, thanks to a combination of measures, including recommissioning thermal power plants that had not been in operation, installing additional diesel power generators, increasing hydropower production by special permission of the Ministry of Land, Infrastructure, Transport and Tourism (which manages rivers), procuring electricity from Chubu Electric Power Co. (which serves a neighboring region), and so on. By July, TEPCO was capable of supplying 57,200 megawatts. During the Tokyo region's hot and humid summers, peak electricity demand usually occurs on summer days when households and offices all use air conditioning at the same time. During past summers, demand during this period peaked at 60,000 megawatts, which caused many people to be concerned about whether or not the region could get through the summer of 2011 without a major power outage.

The accident at Fukushima Daiichi transformed the mentality of citizens, resulting in extensive energy conservation measures in the greater Tokyo area. TEPCO began to provide detailed information indicating the energy currently being used as a percentage of the maximum output capacity of its power plants, and this information was displayed on television, radio, the Internet, display panels in train stations, and so on.

To limit electricity demand during the summer, the government called upon large consumers of electricity to make a 15 percent reduction in electricity consumption, based on legislation, and requested energy saving from all consumers including citizens. Some city governments, supermarkets, and convenience stores offered small gift rewards to citizens who produced evidence from their power utility statements indicating that they had achieved certain energy saving targets. The word *setsuden* (energy conservation) became a key word in the Tokyo area. The brightness of lighting was reduced in subway stations and shopping malls, creating a somewhat darker Tokyo (which had previously been much brighter compared to many cities overseas). There was a surge in sales of LED lamps and a variety of other energy efficient products. As a climate change countermeasure, the government had already been requesting for a number of years that air conditioners be set to keep room temperatures no cooler than 28 degrees Celsius, but this summer, corporations and citizens began to observe this standard more seriously. Government employees wore T-shirts instead of suits as office wear, and businessmen who wore neckties were scolded by clients and suppliers for having poor judgment. Electric fans sold out and were nowhere to be found in stores. Cooperating to reduce peak electricity demand on weekdays, from July to September all auto makers halted operations on Thursdays and Fridays, and operated on Saturdays and Sundays instead.

During the period between the earthquake and the end of August, TEPCO's power usage peaked at 49,220 megawatts, on August 18, 2011, when the maximum temperatures were 36.1 degrees Celsius in Tokyo and 38.3 degrees in the city of Kumagaya, which is known as one of the hottest places in Japan; both cities were then at their hottest temperatures in 2011 (Japan Meteorological Agency 2011). Peak power demand in 2010 was 59,990 megawatts, at 2:00 p.m. on July 23, indicating that conservation measures reduced peak electrical consumption in 2011 by 18.0 percent (TEPCO 2011).

Public awareness had changed. Dentsu Research conducted an Internet survey of 2,000 male and female adults on April 11 and 12, 2011, to which 67.6 percent responded, "I would like to reconsider my former wastefulness and think about what I can save, what I can endure, and what I don't really need to buy," and 52.6 percent responded, "I have become much more conscious about my own use of energy" (Dentsu Communication Institute, Inc. 2011). A Japanese proverb says, "Turn a misfortune into a blessing." These changes in the public's attitude suggest that the potential exists for Japan to move toward being a low carbon society.

To turn this situation into a real blessing, however, an enormous amount of effort is necessary. Replacing nuclear power with a greater share of thermal power generation could simultaneously pose the risk of greater emissions of greenhouse gases (GHGs). Some politicians and mass media commentators, annoyed with the inconvenience of trying to save energy and sweltering in the heat of summer, began to suggest the use of "reserve power." Reserve power refers to power generators installed by corporations for private power generation. They argued that if the sale of electricity from corporations to power utilities were completely liberalized, this power could cover the electricity shortfall resulting from the loss of nuclear power. Much of this reserve power, however, is generated by burning coal. In addition, a growing number of people have been saying that Japan should back out of its medium-term targets to reduce GHG emissions (25 percent reduction by 2020). In response to the Fukushima disaster, Germany and Italy have gone the route of abandoning nuclear power. Is it possible for Japan to choose the path of creating a low carbon society without excessive dependence on nuclear power to generate electricity? The issue for the next few years will be whether or not Japan can truly "turn a misfortune into a blessing."

Challenges for Asia

To achieve the target of limiting the global temperature increase to 2 degrees Celsius, as agreed at the COP16 Climate Conference (i.e., the Sixteenth Session of the Conference of the Parties to the UN Framework Convention on Climate Change, or UNFCCC) in Cancun, the world as a whole must take on the challenge of making the transition toward a low carbon society. The Asian region's impact on the global climate is enormous. This region accounts for 60 percent of the world's population. Its share of carbon dioxide emissions from fuel grew from 23 percent in 1990 to 28 percent in 2008. Per capita emissions from three Asian non-Annex I countries – Brunei, Korea, and Singapore – exceeded those of Japan in 2008 (IEA 2010). Bangkok's 7.1 tons per capita carbon dioxide emission exceeded Tokyo's 5.7 tons per capita by 25 percent in 2005 (Bangkok Metropolitan Administration 2007). It is clear that the issue of whether or not it will be possible to limit carbon dioxide emissions from Asia will dramatically affect the future of the global environment.

In common with a separate book on adaptation to climate change published in 2010 by the JICA Research Institute (Fujikura and Kawanishi 2010), this book is a summary of knowledge about climate change mitigation for practitioners involved in countermeasures in developing country governments, development assistance organizations, non-governmental organizations, and so on. The 2010 book mainly summarized lessons learned from projects implemented in Asia and Africa. The larger number of examples from Africa reflected that continent's vulnerability and need for climate change adaptation. This current book, on the other hand, introduces examples from Asia that are important from the perspective of climate change mitigation.

This book has two parts. Part I reviews the current status of, and challenges for creating, an international framework on mitigation. It begins by reviewing and analyzing the UNFCCC and COP climate conferences, then the Clean Development Mechanism (CDM) and REDD+, and concludes by covering international intercity collaboration. Part II introduces examples from developing countries. It begins by analyzing policies in China and India, then provides examples of nationally appropriate mitigation actions (NAMAs) in Thailand's waste and wastewater management sector, discusses policy support to tackle climate change in Indonesia and Vietnam, as well as the GHG reduction effects of building efficient coal-fired thermal power plants in the Asian region and photovoltaic power generation in Bangladesh, and ends by introducing initiatives to create a roadmap for moving toward a low carbon society in Japan. Finally, the book includes a commentary contributed by the Environmental Law Institute, an American NGO composed of environmental law professionals.

The stance of developing countries

In Chapter 2, Yasuko Kameyama reviews the history of related discussions, from the creation of the Intergovernmental Panel on Climate Change (IPCC) in 1988 to the COP17/CMP7 Climate Conference in Durban, South Africa, in 2011. Under the Bali Action Plan adopted in 2007 it was agreed that final agreement on a framework to succeed the Kyoto Protocol would be reached at COP15 in 2009. It was not made clear, however, whether the aim was a new protocol, a revision of the Framework Convention, or just decisions at a series of COP meetings. At the same time, the Ad hoc Working Group on Further Commitments for Annex I Parties under the Kyoto Protocol (AWG-KP), which debates the post-commitment period relating to Article 3.9 of the Kyoto Protocol, carried out discussions parallel to COP, but on the assumption that the Kyoto Protocol would be extended. When

COP15 was held in Copenhagen, the relationship between COP meetings and the AWG-KP had still not been clarified. The COP conference was not able to achieve the objectives of the Bali Action Plan, but was only able to "take note" of certain points in the Copenhagen Accord.

The general view is that almost nothing was achieved at COP16 in Cancun. Nevertheless, the Cancun Agreements were adopted, thanks to an initiative by host country Mexico. The content of the Cancun Agreements was about the same as that of the Copenhagen Accord, but the significance of the Agreements was that the Cancun meeting took what had simply been an expression of political agreement and adopted a text that linked it to the formal UNFCCC process.

Kameyama points out that developing countries, while avoiding international commitments, have a number of motives to implement mitigation measures. First, developing countries, like other countries, also have inadequate supplies of energy and need to achieve energy security. Moreover, environmental problems arising from energy production and consumption are becoming noticeable domestically in these countries. Participating actively in international negotiations relating to climate change is also meaningful from the diplomatic perspective, and in order to obtain financial support. Furthermore, the introduction of low carbon policies can contribute to a nation's industrial development. In the context of regional frameworks like the Asia Pacific Economic Cooperation (APEC) forum, developing countries have been proactive in actions outside the COP framework, such as with the declaration on climate change and energy adopted in 2007. Developing countries also have considerable potential to promote initiatives at the regional, national, local, and citizen levels, and Kameyama anticipates that progress at these levels will occur more rapidly than at the international level, which has been slow to date.

Clean Development Mechanism

The CDM is one of the Kyoto Mechanisms, together with Joint Implementation and emissions trading. One of the CDM's features is that it targets non-Annex I countries and includes the objective of contributing to sustainable development in developing countries. To date, 2,800 projects have been registered by the CDM Executive Board as CDM projects, and 5.5 million Certified Emission Reduction (CER) units have been issued. The CDM has unquestionably contributed to measures to address climate change, but various problems with it have been highlighted, such as the long amount of time required for a project to get from the application to the registration stage, and the fact that typical energy conservation projects cannot easily obtain registration as CDM projects.

In Chapter 3, Yamada and Fujimori analyze 2,500 of the CDM projects registered to date, and note regional and sectoral imbalances. For 41 percent of registered projects, China is the host country, and China, India, and Brazil together account for 70 percent of the total. Of the total CER credits issued, 51 percent were for decomposition of HFC-23 and 13 percent for decomposition of nitrous oxide (N_2O). In effect, the CDM is concentrated on the decomposition of GHGs other than carbon dioxide in super-developing countries where implementation is easy and investment is efficient. As a result, even though contributing to sustainable development in developing countries is one of the objectives of the CDM, one would have a hard time saying that this is being adequately achieved.

To improve the CDM, Yamada and Fujimori point out that, to begin with, the current registration process is complex and needs to be simplified. Regarding CDM project registration, they propose adopting a bottom-up approach with sustainable development kept clearly in mind, and incorporating other benefits from projects (i.e., co-benefits) besides the reduction of GHG emissions only. In addition, they say that it is necessary to create mechanisms to ensure financing for projects that have been registered as CDM projects.

REDD+

In Chapter 4, Mitsuo Matsumoto gives an overview of REDD+, which is a new mechanism that has generated high expectations. The term "REDD+" stands for "Reducing Emissions from Deforestation and forest Degradation and the role of conservation, sustainable management of forests, and enhancement of forest carbon stocks in developing countries." The REDD component is the reduction of emissions caused by forest loss and degradation, and the "+" (also spelled out as "Plus") refers to maintaining and increasing GHG sequestration after forest conservation has been implemented. The CDM also covers reforestation and afforestation, but REDD+ addresses forest conservation activities more broadly than the CDM does. If REDD+ is implemented, funding will be provided for forest conservation in developing countries, and it can be expected to contribute to local community development and conservation of biodiversity.

Discussions on REDD+ began at the COP11 Climate Conference in Montreal in 2005. It is also referred to in the Cancun Agreements as "enhanced action on mitigation." REDD+ promotes voluntary private sector initiatives as well as multilateral bilateral initiatives involving international agencies such as the World Bank, Food and Agriculture Organization (FAO), and United Nations Development Programme (UNDP). However, because discussions at COP meetings have made little progress, REDD+ is not yet recognized as a program under the UNFCCC. For that to happen, it is first necessary to settle issues regarding methodologies. Use of a combination of remote sensing and ground surveys has been agreed upon to estimate the amount of change in forest carbon stocks, but specific methodologies have not yet been worked out. Furthermore, in order to create economic incentives, measurement, reporting, and verification are essential, for which methodologies must be established. As for funding sources, various proposals have been made, including the creation of a fund, the use of market credits, and a staged process starting with international funding and ultimately using market credits. Also, even if REDD+ methodologies were to be settled on, the program would be meaningless unless policies were in place in the countries concerned to implement them properly. Measures to deal with such policy concerns are also important.

Intercity collaboration

Local governments play an important role in the movement toward low carbon societies. This is not only because local governments are able to promote awareness raising, and provide guidance to citizens and local companies, but also because they have authority over building and housing construction permits, as well as transportation policy and urban planning. Japanese local governments are also offering their knowledge of low carbon society policies and technologies to local governments in Asia.

Kitakyushu City, in Japan, is an example of a local government actively engaged in international collaboration. This city has declared a target of reducing carbon dioxide emissions in the city by 50 percent from 2005 levels by 2050, and also of reducing emissions in the Asian region by an amount equivalent to 150 percent of the city's emissions. The Kitakyushu Asian Center for Low Carbon Society has been established as a core facility for achieving these targets while also promoting local economic development and is providing technical support, information, and more, to cities in Asia (Kitakyushu 2011).

Chapter 5 by Hidenori Nakamura and Hideyuki Mori summarizes the status to date of international intercity collaboration on environmental issues in Asia, and highlights the need to promote a platform to encourage collaboration between Japan and Asian cities for advancing low carbon development. They investigate four international intercity networks (Kitakyushu Initiative for a Clean Environment, the Clean Air Initiative for Asian Cities, ICLEI, and CITYNET), as well as the current status of the bilateral intercity environmental collaboration deriving from these networks. Collaborative projects have been conducted in the past to address specific issues, such as environmental education, waste treatment, and recycling; based on the experience gained, it will be possible to promote intercity collaboration directed toward low carbon development. For example, in Japan, voluntary carbon offsets are being promoted between large cities and rural areas, but the potential also exists to promote intercity technical cooperation and business development for local companies to promote low carbon development. Generally speaking, however, local governments are not strongly motivated to involve themselves in international collaboration, and they have limited personnel capacity and financial resources to dedicate to such efforts. The experience and know-how needed for international collaboration is lacking. To resolve such issues, Nakamura and Mori propose the creation of an international platform on low carbon cities to connect cities in Japan and Asia. Because local governments lack abundant resources, such a platform could contribute to low carbon development in Asian cities if the Japanese government were to provide financial support, and the platform's secretariat could play a coordinating role.

Policies of China and India

In Chapter 6, Frank Hiroshi Ling analyzes four sectors (energy, industry, transportation, and building) in terms of mitigation measures in China and India. These two countries are the world's first- and fourth-largest emitters of greenhouse gases, respectively; with respective populations of 1.3 billion and 1.1 billion, their emissions are rising rapidly with economic growth. They account for more than 60 percent of the increase in global GHG emissions from 2000 to 2007. Both countries emphasize "historical responsibility" for climate change, and have been taking a cautious attitude toward adopting mandatory emission reductions. Nonetheless, they have announced a commitment to reduce emissions per unit of GDP in response to the Copenhagen Accord. Energy consumption is increasing rapidly in both countries, and in the long term it is uncertain if the international market would be able to supply enough fossil fuel to meet their demands. Raising efficiency is also an important issue for both countries from the perspective of national energy security. Pollution from power generation, industry, and transportation is becoming an important consideration, so the possible co-benefits from measures to reduce greenhouse gases are also significant. China has met its target of a 20 percent reduction in energy consumption per unit of GDP established under the country's Eleventh Five-Year Plan. India developed its first National Action Plan on Climate Change in 2008.

Chinese and Indian corporations are actively promoting renewable energy projects. In 2009, four of the world's top ten manufacturers of solar cells, ranked by size of production, were Chinese companies. That same year China ranked second in installed wind power generation capacity, and India ranked fifth (NEDO 2010). Nevertheless, both countries are still heavily dependent on fossil fuels, and this dependency is generally not expected to change in the future. It is important for them to decommission inefficient, aging coal-fired thermal power plants and introduce highly efficient ones. China's electricity demand is expected to rise rapidly; meanwhile, in India, 600 million people do not yet have access to electricity. Under these circumstances, Ling points out that there is an opportunity to "leapfrog" over conventional carbon-intensive patterns of development by creating smart grids. To cover the costs of low carbon development, China and India will reportedly need to invest 2.8 trillion US dollars and 1.0 trillion US dollars by 2030, only in the power sector. Support in terms of funding by the international community will be important, but in order to facilitate procuring private sector funds, structural reforms are also necessary.

Nationally appropriate mitigation actions (NAMAs)

NAMAs, which are to be carried out by developing countries, were formally adopted in the Bali Action Plan. At COP16 in Cancun, in the context of sustainable development, it was agreed that developing countries would receive technological and economic assistance to implement NAMAs. It was decided that mitigation measures conducted with international support would be in accordance with guidelines established based on the UNFCCC, and be implemented in such a manner that they would be measurable, reportable, and verifiable (MRV). However, a consensus was not reached among countries regarding the modalities and procedures for NAMAs and monitoring, reporting, and verification systems. In Chapter 7, Kazuhito Yamada and Mariko Fujimori present a case study on the waste treatment sector in Thailand, and propose a concrete approach for how to proceed in the case of implementing NAMAs.

They describe key issues that should be considered when implementing NAMAs in an MRV manner. First, NAMAs should not only reduce greenhouse gas emissions but also contribute to the sustainable development of the countries concerned. Thus, the first task is to identify problems in the sectors where NAMAs are being implemented, and then to consider how to reduce greenhouse gas emissions through countermeasures that are identified in the first task. To this end, they first collected information about various types of policies and measures relating to waste, for the Thai government and also for the city of Bangkok, and then identified problems. The second key issue is then how to establish MRV systems. Yamada and Fujimori considered CDM methodologies and formulated reference scenarios for each task relating to waste treatment. Based on that information, they established target values for greenhouse gas emissions. Next, they prioritized NAMA projects, and established MRV methodologies for each activity. The third key issue is how to achieve financing of NAMAs. Here, they considered the most appropriate funding source for each activity, from sources such as the Thai national government, the Japan Bank for International Cooperation (JBIC), private sector companies, and so on. The final key issue is to share this NAMA experience with other developing countries, which would be facilitated by international meetings, training, and seminars relating to NAMAs and MRV. Aiming to find ways to implement NAMAs in an MRV manner, they conclude with questions that remain regarding how to quantify indirect reductions of greenhouse gas emissions that are achieved in implementing projects such as institutional reforms, awareness-raising, training, and so on.

Development policy lending

In Chapter 8, Taro Katsurai and Yuka Murakami review development policy lending (DPL) being implemented in Indonesia and Vietnam as climate change countermeasures. DPL is a type of policy-based lending instrument, and is provided with the objective of extending financial assistance to developing country governments. The grant conditions require the developing country governments to prepare policy reforms with a focus on sustainable development, through discussions with donors, and then to implement these reforms.

In the lead-up to the COP13 Bali Conference in 2007, the Indonesian government prepared a National Action Plan on Climate Change (NAP-CC), and the document "National Development Planning: Indonesian Responses to Climate

Change" (also known as the "Yellow Book"), as national actions of its own to address climate change. In order to facilitate implementing these actions, a Climate Change Program Loan was developed. The conditions for extending this loan required the Indonesian government to prepare a policy matrix consisting of three pillars (mitigation, adaptation, cross-cutting issues), and to commit to implementing them from 2007 to 2009. JBIC (currently the Japan International Cooperation Agency, JICA) agreed to the loan and extended 300 million US dollars to Indonesia in 2008. As co-financer, Agence Française de Développement (AFD) also extended 200 million US dollars to Indonesia the same year. These funds were used by the Indonesian government to cover a fiscal deficit. In 2009, it was judged that implementation of the policy matrix had been satisfactory, so JICA and AFD paid out an additional 400 million and 300 million US dollars, respectively.

Vietnam formulated a National Target Program to Respond to Climate Change (NTP-RCC) in 2008. It indicated 26 activities that should be implemented by 2015, and estimated that this would require about 110 million US dollars. In order to implement the NTP-RCC, the Vietnam government conducted intensive discussions with many participating stakeholders, and in 2009 formulated its Support Program to Respond to Climate Change (SP-RCC) with a focus on institution building. Having confirmed these results, JICA extended about 120 million US dollars, and AFD 28 million US dollars, to the government of Vietnam.

Unlike project financing, DPL funds are extended in one single payment to a national financial authority. The advantage of this approach is that it is unnecessary to calculate the amount of funding required for each activity, and the overall processing effort is reduced. However, because the funds are provided only to the national financial authority, there is a need to motivate the government agencies concerned to actually implement the activities listed in the policy matrix. Developing country governments are serious about making international commitments to achieve greenhouse gas emission reduction targets, but in the case of DPL, rather than aiming for numerical targets, they can obtain funds by introducing appropriate policies; the advantage here is that both donor and developing country governments can more easily accept the details.

Efficient coal-fired thermal power generation

The electrical power sector is the largest source of carbon dioxide emissions, and the increase in global carbon dioxide emissions is largely driven by the power sector. In 1971 the power sector accounted for 25 percent of global carbon dioxide emissions; by 2000 it had increased to 35 percent (Chapter 10). Coal is widely used as a fuel for electricity generation, with more than 40 percent of electricity generated at coal-fired power plants. Coal is a dirty fuel with the highest carbon intensity among fossil fuels, but because it is relatively abundant and low cost, unless there is a major change in the price of coal, the amount of electricity produced by coal-fired power plants is expected to double from the 2007 level by 2035. An important challenge is to limit the amount of carbon dioxide generated from coal-fired power plants in developing regions. The introduction of carbon capture and storage (CCS) technology is being considered for coal-fired power plants, but the prospects for practical applications of it are still far away due to the economic and technical factors involved. For the foreseeable future, the option that remains is to increase the efficiency of coal-fired electricity generation.

Since the 1990s, Japan has been providing loans under official development assistance (ODA) for coal-fired power plant construction projects in four countries: China, India, Malaysia, and Vietnam. Advanced technologies were used in the power plants built with these yen loans, and they now generate electricity with greater thermal efficiency than existing power plants. To increase their supplies of electrical power, developing countries can better limit carbon dioxide emissions by using these yen loans to introduce advanced technologies and to build new leading-edge coal-fired plants, rather than by expanding the capacity of existing plants and building new conventional power plants. In Chapter 9, Tomoyo Toyota establishes four reference scenarios based on CDM estimation methods to evaluate the carbon dioxide emission reductions from coal-fired power plant projects implemented at ten locations in four countries using ODA yen loans. She finds that at the ten plants (total power generation capacity 7,180 megawatts), carbon dioxide emissions were reduced by between 4.1 million and 7.5 million tons, compared to the reference scenarios, for the period 2000 to 2007. This amounted to between 0.4 percent and 0.8 percent of Japan's carbon dioxide emissions during the same period.

Solar home systems

In Chapter 10, Shinji Kaneko, Satoru Komatsu, and Partha Pratim Ghosh analyze the electrical power sector in Bangladesh. They find that electrification in Bangladesh is still limited, with only 28 percent of rural households having electrical power in 2008. Grid electrification is being funded mainly by the World Bank, Asian Development Bank, and JICA. At the rural level, the Rural Electrification Board (REB) and *Palli Bidyut Samity* (PBS), a business entity under its authority, are major actors in promoting electrification. Following a system carefully designed to prevent corruption and political involvement, they are promoting electrification and fee collection. Households under contract with the REB rose from one million in 1995 to more than six million in 2009. The fee collection rate is steady at nearly 100 percent. However, as electrification continues, the growth in generation capacity cannot keep up with the growing demand. Peak demand in 2009 was 6,066 megawatts, whereas the supply was only about 70 percent of that amount, at 4,162 megawatts.

Solar home systems (SHSs) are rapidly spreading in rural areas that are not yet electrified. These systems are expensive, costing from 300 to 1,000 US dollars, but with support from the Infrastructure Development Company Limited (IDCOL) to NGOs promoting the systems, installations are occurring, starting with relatively more affluent households in rural areas. In 2010, 763,957 sets were installed, a number that exceeded expectations. The funders of this program include the World

Bank (IDA), GTZ/KfW, the ADB, the Islamic Development Bank (IDB), the Global Partnership on Output-Based Aid (GPOBA), and other agencies.

The major motives for installing the solar home systems are for children's home study, to watch television, and to charge mobile phone batteries. The majority of people living in non-electrified areas charge their mobile phones at the market or at the house of a neighbor or friend; the greater the number of telephones in a household, and the farther the household is from a location where a battery can be charged, the greater the motivation to install a solar home system. Households that own many batteries, and those that use a lot of kerosene for lighting, are also more likely to introduce a solar home system.

Of the 20 million households in rural Bangladesh, only 5.5 million are electrified. Even if 600,000 households install solar home systems, electricity will still not be available in about 14 million households. The obstacle preventing the further spread of solar home systems is cost. The National Energy Development Organization (NEDO) in Japan is aiming for a cost breakthrough in photovoltaic power generation through technological innovation, and has established a plan to bring the cost of electricity generation down to 7 yen per kilowatt-hour, significantly below the current market price of electricity at 23 yen per kilowatt-hour. If this target is reached, the spread of the technology will be given a big impetus not only in Japan but in developing countries as well. Development cooperation has hitherto focused on the transfer of proven technologies, but to facilitate the transfer of new technologies in the future, it will be important to promote further collaboration among technology research and development organizations and development assistance organizations.

Making Japan a low carbon country

In response to Cool Earth 50 (a strategy proposed by then Prime Minister Shinzo Abe in 2007), the government of Japan adopted the Action Plan for Achieving a Low-Carbon Society by Cabinet decision in July 2008, and declared its intention to reduce emissions from current levels by 60–80 percent in the year 2050 as the long-term target for Japan. The basis for this government decision was a set of scenarios for realizing a low carbon society, developed mainly by the National Institute for Environmental Studies (NIES) and Kyoto University. In Chapter 11, Junichi Fujino, a member of the development team, explains the history of this initiative up to the Japanese government's adoption of the low carbon policies based on the scenarios the team developed.

The research team also created a roadmap for cutting the greenhouse gas emissions of Shiga Prefecture in half by 2030 from 1990 levels. The roadmap showed the cost burden to achieve this target, and in February 2011 the Prefectural Assembly passed the Shiga Prefecture Ordinance on Creation of a Low Carbon Society, which was based on the roadmap. The research team also found that it would be technically possible for Kyoto City to reduce its carbon dioxide emissions in 2030 by 40 percent or more compared to 1990 levels. In September 2010, the Kyoto City Council passed, by consensus, an amendment to the city's Climate Change Ordinance, declaring emission reduction targets of 25 percent by 2020 and 40 percent by 2030.

Moreover, in cooperation with the Universiti Technologie Malaysia, the research team prepared a low carbon scenario for Malaysia's Iskandar City. The team predicted that under a business-as-usual scenario, greenhouse gas emissions would rise by 262 percent from 2005 to 2025, but if low carbon policies were implemented, the increase could be limited to only 52 percent.

These were momentous initiatives by the Government of Japan, with Shiga Prefecture and Kyoto City, to formulate measures through collaboration among the central government, local governments, and a research institute, and then to have these measures adopted. It is important, however, to work at consensus building among stakeholders when implementing these programs. A workshop will be implemented with stakeholders in Malaysia, although the involvement of experts is still not sufficient. To realize low carbon societies, consensus building among experts, citizens, and other stakeholders will be a future challenge.

Commentary on Parts I and II

We invited the Environmental Law Institute to provide a commentary on Parts I and II. As the title of their commentary, "Not Such Strange Bedfellows: Climate Mitigation as a Buttress for Sustainable Development," implies, Nichols, Pendergrass, and Connolly stress that economic development and mitigation should be pursued in tandem in developing countries, because their main priority is economic development. In order for developing countries to conduct substantial mitigation measures, they recommend that a "policy-based approach," by which developing countries conduct mitigation measures on their own initiative, has to be adopted, because these countries have rejected a "target-based approach." From this perspective, DPL, as reported in Chapter 8, could be an effective tool. Nichols, Pendergrass, and Connolly go on to argue that developing countries will only conduct mitigation measures if they include appropriate co-benefits. REDD+, reported in Chapter 3, and NAMAs in the waste management sector, reported in Chapter 7, could play an important role for developing countries to receive co-benefits.

Nichols, Pendergrass, and Connolly recognize the necessity of improving the CDM in order for more relevant projects to be registered, as recommended in Chapter 3. They also emphasize that both financial and technical support for formulating CDM projects should be better provided for developing countries. In order to establish an appropriate MRV methodology, it is crucial to find a way to evaluate indirect GHG reductions, such as those that occur through institutional and educational activities. As long as developing countries have to depend on coal-fired power stations, it could be appropriate to recognize establishing highly efficient coal-fired power plants, such as those reported in Chapter 9, through NAMA, and/or the CDM. Moreover, taking account of stagnant international policy formulation, Nichols, Pendergrass, and Connolly emphasize the importance of intercity cooperation on mitigation, as reported in Chapter 5.

They also argue for the effectiveness of a sectoral approach. Because of time constraints, we conducted case studies only on the energy and waste management sectors. Nonetheless, measures in other sectors such as transportation, housing, and urban planning are also important. Studies on sectors other than energy will be subjects of our future studies.

Conclusion

Nineteen years have passed since the adoption of the UN Framework Convention on Climate Change, and 14 since the Kyoto Protocol was adopted. During this time, scientific knowledge has accumulated, and concerns about the impacts of climate change are more deeply shared. Actual impacts have started to become noticeable, and some regions are being forced to take adaptation measures. This has also changed the stance of developing countries, which was initially negative with regard to implementing climate change countermeasures. Countries are now seriously debating climate change countermeasures, and some have been adopted through the Copenhagen Accord and Cancun Agreements, and so on.

Developing countries today are still taking a cautious attitude toward making international commitments to reduce their greenhouse gas emissions, but they are motivated to reduce emissions. If they limit their emissions, they can also expect to reduce their energy costs, improve energy security, and reduce environmental pollution arising from energy use. Technological advances are also being made toward energy conservation and a low carbon society, and Asia has become an enormous market for these technologies. Asian companies not only import technologies from developed countries, but also sell their own technologies and products. China has become the world's top producer of photovoltaic panels. Companies from Asian countries are also participating in research and development of highly efficient coal-fired power plants.

In fact, Asian countries are in the process of reducing their amount of greenhouse gas emissions per unit of GDP. The problem is that these efforts alone would be inadequate if they were conducted alongside business as usual. Separate from the issue of whether or not international commitments are made, it will be impossible for humanity to limit the global temperature increase to the 2 degrees Celsius agreed to in Cancun unless Asia increases its current efforts. Initiatives to strengthen the international framework to that end are moving ahead in various forums. Such efforts include REDD+ and the formulation of NAMAs that can be measured, reported, and verified. Governments, research institutions, NGOs, and corporations need to actively make further efforts like these. Progress with international intercity collaboration is also promising as a new platform. Local governments occupy a special position closest to the activities of citizens, and play an important role in terms of promoting grassroots climate change countermeasures. Intercity sharing of experiences among citizens and local governments is very valuable, as they are based on real experience, even if they may not be large or dramatic.

Whether developing countries can leapfrog from a business-as-usual situation to a low carbon society depends on whether or not they can procure the required funds and technology. Technologies for wind and solar power are already commonly available in Asian countries. If costs are reduced due to competition and technological innovation, new forms of energy will spread more quickly. It is not only corporations that can play a role in spreading new technologies in developing countries, but also donors and NGOs.

Even if new forms of energy were to spread rapidly, each country's dependence on coal would probably not drop very quickly in the foreseeable future. Thus, one strategy to address climate change would be for donors to support the spread of highly efficient coal-use technologies. At present such projects probably could not be registered as CDM projects, but it is still important to think about how such activities can be encouraged internationally, and how to provide incentives to donors.

Through development policy lending and the like, with a focus on climate change policies, if it is possible to actively incorporate the low carbon perspective of NAMAs into development strategies, funding will contribute to energy conservation and new energy market expansion in the countries concerned. While becoming more active in supporting initiatives such as development policy lending that addresses climate change, and so on, donors would likely promote initiatives to share lessons learned and outcomes among other donors and among developing countries.

Returning now to Japan, the current situation is that, since the Fukushima nuclear disaster, throughout Japan it has become very difficult politically to restart nuclear power plants that had been shut down for regular maintenance. Local governments have become more concerned, and citizens more reluctant to see them restarted, than they were before the disaster, or have even switched over from accepting to rejecting nuclear power. With the exception of Okinawa Prefecture, which does not have nuclear power plants, all of Japan is facing the risk of power shortages. The temptation to fall back on fossil fuel dependence is rising, but instead Japan should make the jump toward becoming a low carbon society, using this crisis as an opportunity. Until now Japan has been transferring and exporting energy efficient technologies to developing countries through official development assistance and private sector cooperation. Rising above today's difficulties and transferring this experience to developing countries can be Japan's contribution to the world, and a return of gratitude to friends who have helped Japan.

There are limits to the funding and technologies that donors and NGOs can provide. However, by using them effectively, donors and NGOs can support the efforts of the developing countries themselves, and can play a role in leveraging more private-sector funding for low carbon development. This book has introduced just a small part of that potential. The specific initiatives mentioned were limited to the power and waste sectors, as time and space limitations prevented coverage of the transportation sector, urban planning, the housing sector, and so on – also important sectors for mitigation measures. Also, many roles for innovative and effective donors could not be covered in this book. I continue to look forward to the day when the entire world makes the jump toward being a low carbon society, by sharing and integrating various low carbon development initiatives by NGOs working on the ground doing development cooperation, as well as by donor organizations, developing country governments, developed country governments, local governments, researchers, private corporations, local communities, ordinary citizens, and so on.

Acknowledgments

I would like to thank all the dedicated contributors who submitted high quality papers and commentaries for this book. I also appreciate advice received in the preparation of this book from Mr. Ichiro Adachi, Mr. Yasuhiko Sato, Ms. Yoshiko Isozaki, and Ms. Yukiko Aida of the JICA Research Institute. I also wish to acknowledge the many valuable comments received from staff at JICA Headquarters.

References

Bangkok Metropolitan Administration (2007) *Action on Global Warming Mitigation 2007–2012*. Bangkok: Bangkok Metropolitan Administration.

Dentsu Communication Institute, Inc. (2011) Predictions of Change in "Awareness, Lifestyles, Societal Systems" to Support Japan's Regeneration. Dentsu Communication Institute, Inc., Tokyo.

Fujikura, Ryo and Masato Kawanishi (Eds.) (2010) *Climate Change Adaptation and International Development*. London: Earthscan.

International Energy Agency (IEA) (2010) CO2 Emissions from Fuel Combustion: Highlights. Online, available at: www.iea.org/co2highlights/co2highlights.pdf (accessed 29 August 2011).

Japan Meteorological Agency (2011) Historical Weather Data. Online, available at: www.data.jma.go.jp/obd/stats/etrn/index.php (accessed 20 September 2011).

Kitakyushu City (2011) Kitakyushu Asian Center for Low Carbon Society. Online, available at: http://asiangreencamp.net/eng/ (accessed 26 September 2011).

Ministry of Economy, Trade and Industry (2011) Future Electricity Supply Forecasts and Responses from Tokyo Electric Power Corp Facilities. Press release, March 25.

National Institute for Environmental Studies (NIES) (2011) Low Carbon Society Scenarios towards 2050. Online, available at: http://2050.nies.go.jp/index.html (accessed 26 September 2011).

New Energy and Technology Development Organization (NEDO) (2010) NEDO White Paper on Renewable Energy. NEDO, Kawasaki.

Tokyo Electric Power Company (2011) Historical Data on Actual Electrical Usage. Online, available at: www.tepco.co.jp.cache.yimg.jp/forecast/html/download-j.html (accessed 29 August 2011).

PART I

International framework on mitigation

2

A REVIEW OF DISCUSSIONS ON AN INTERNATIONAL FRAMEWORK TO ADDRESS CLIMATE CHANGE, AND MITIGATION ACTIONS BY DEVELOPING COUNTRIES

Yasuko Kameyama

Introduction

Climate change affects everyone and everything on earth – not only human beings, but also other species and the entire ecosystem. The adverse impact of climate change will bring the most harm to those who are the most vulnerable. This does not necessarily mean that the most vulnerable are the largest emitters of greenhouse gases (GHGs). Rather, in many cases it is the other way around. Industrialized countries emitted most of the GHGs in the past, and they are now wealthy enough to be able to cope, to a certain extent, with the adverse impact of climate change. This also explains why, when the United Nations Framework Convention on Climate Change (UNFCCC) was adopted in 1992, the industrialized nations agreed that they would start to reduce GHG emissions before other countries did so, and made a commitment to provide financial as well as technological assistance to less developed countries.

While industrialized countries struggle to reduce their own GHG emissions, members of the group of developing countries experiencing economic growth, namely the G77 plus China group, have followed a more diverse trajectory within the last decade. Countries such as Mexico and Korea joined the Organization for Economic Cooperation and Development (OECD) in 1994 and 1996, respectively, which sent the message that they had reached a relatively wealthy status. Brazil, South Africa, India, and China have formed the BASIC group to coordinate their positions at international negotiating meetings on climate change policies. They share similar national circumstances, as their emissions per capita have been climbing due to their rapidly growing economies and rising consumption of energy. Although their per capita emissions remain relatively low compared to those of industrialized countries, the absolute amount of GHG emissions in these countries now has a substantial impact. China's GHG emissions are said to have surpassed those of the United States at some point in 2007. Today it is the largest emitter, and accounts for nearly 22 percent of global carbon dioxide (CO_2) emissions.

Although industrialized countries still bear the most responsibility, due to their cumulative historical emissions, and should take the lead in emissions reduction, it has become difficult to deny the increasing importance of emission mitigation actions taken by developing countries. This chapter explains the history of multilateral cooperative action to tackle climate change, focusing on issues related to mitigation activities in developing countries. The UNFCCC has taken on the central role in the multilateral process, but climate change has also been addressed by other international organizations and multilateral forums. The latter part of this chapter reviews some of the regional or bilateral channels that provide climate mitigation actions. The concluding section summarizes ways to move forward.

History of climate change negotiations under the UNFCCC

Awareness that climate change is a problem became more widespread internationally in the late 1980s (Table 2.1) The Intergovernmental Panel on Climate Change (IPCC) was established in 1988 to compile scientific findings related to climate change. Its first assessment report came out in 1990, and stressed the great risk of climate change in the future. Countries agreed to start negotiations under the United Nations to deal with climate change at a multilateral level, and reached an agreement in May 1992. That agreement was called the United Nations Framework Convention on Climate Change (UNFCCC) (Mintzer and Leonard 1994).

Under the UNFCCC, all countries, including developing countries, made a commitment to submit emissions inventories and national communications (national

TABLE 2.1 History of climate change negotiations

Year	*Developments in international negotiations on climate change*
1988	IPCC established
1990	The First IPCC Assessment Report
1991–1992	International negotiations on climate change agreement Adoption of UNFCCC
1994	UNFCCC entered into force
1995	COP1, Berlin Mandate adopted
1997	COP3, Kyoto Protocol adopted
2001	Withdrawal of the United States from the Kyoto Protocol COP7, Marrakesh Accords adopted
2005	Kyoto Protocol entered into force COP11, CMP1
2007	COP13, CMP3, Bali Action Plan adopted
2009	COP15, CMP5, Copenhagen Accord adopted
2010	COP16, CMP6, Cancun Agreements adopted
2011	COP17, CMP7

Source: author.

reports on implementation of the Convention) to the Secretariat (Article 4.1). In addition, industrialized countries and economy-in-transition countries such as Russia, which are called Annex I countries, committed to implement climate mitigation policies aimed at stabilizing GHG emissions at 1990 levels by 2000 (Article 4.2). Furthermore, the industrialized countries, known as the Annex II countries, were obliged to provide "new and additional financial resources to meet the agreed full costs incurred by developing countries in complying with their obligations" to report their national inventories and communications (Article 4.3). They were also requested to "take all practicable steps to promote, facilitate and finance, as appropriate, the transfer of, or access to, environmentally sound technologies and know-how" to developing countries (Article 4.5).

However, the treaty did not require developing countries to make any commitment on the amount of emissions stipulated in the UNFCCC. The negotiators at the time were aware of there being a wide gap between the developed and the developing countries in terms of their levels of responsibility as well as their capability to reduce GHG emissions (Agarwal 2002). The key phrase "common but differentiated responsibilities and respective capabilities" was inserted in Article 3.1 of the UNFCCC to describe this situation. It was also stated that

> the extent to which developing country Parties will effectively implement their commitments under the Convention will depend on the effective implementation by developed countries of their commitments under the Convention related to financial resources and transfer of technology and will take fully into account that economic and social development and poverty eradication are the first and overriding priorities of the developing country Parties.
>
> *(Article 4.7)*

The UNFCCC entered into force in 1994. The first Conference of the Parties to the UNFCCC (COP1) was held in Berlin in 1995. The participants had no objection to stating that commitments under the UNFCCC were not adequate to mitigate climate change below the level that would prevent a dangerous effect on ecosystems. At the same time, countries held diverse views when it came to discussing what should be done next to improve the adequacy of commitments. Industrialized countries pointed at developing countries' rapidly growing emissions, saying it was now time for developing countries to commit to emission mitigation actions. Developing countries pointed to industrialized countries, accusing many of not even trying to achieve the emission target set by Article 4.2(a) and (b) of the UNFCCC, which committed the developed countries to stabilizing emissions at 1990 levels by 2000. After long and intense negotiations, an agreement called the Berlin Mandate was reached to start a new round of negotiations to set emission reduction targets for Annex I countries (UNFCCC 1995). The mandate clearly stated that the new legal instrument would not introduce any new commitments for developing countries, but would reaffirm existing commitments in order to achieve sustainable development, taking into account articles in the UNFCCC related to financial and technological assistance. This statement was inserted because of the insistence by the developing countries' group that emissions in many developed nations had continued to rise even after they had ratified the UNFCCC. The developing countries demanded that industrialized countries' emissions be reduced first before developing countries would agree to take any action to mitigate emissions.

After two years of difficult negotiations, the Kyoto Protocol was adopted in Kyoto in the final hour of COP3 in December 1997 (Grubb *et al.* 1999; Oberthür and Ott 1999).

Article 3.1 of the Kyoto Protocol indicated Annex I countries' commitment to reduce or limit their emissions to the specific levels stipulated in Annex B of the Protocol. The countries were allowed to count emission sequestration in forests and from land-use changes, as well as being allowed to utilize carbon markets, i.e., emissions trading, joint implementation, and Clean Development Mechanisms (CDM). Countries continued work to determine detailed rules for implementing the Protocol's agreements, and reached an agreement at COP7 in 2001, known as the Marrakesh Accords (UNFCCC 2002).

The Kyoto Protocol entered into force several years later, in February 2005. The first COP that served as a meeting of the Kyoto Protocol (CMP) was held in December 2005 in Montreal. The participants in COP and CMP were almost the same, except for the United States, which was a party to the UNFCCC but an observer for the Kyoto Protocol. The two-track process was inevitable as there were two independent international agreements, but it was a cause of confusion for many.

The Kyoto Protocol had set emission reduction targets only for Annex I Parties for the years between 2008 and 2012. This meant that a new round of negotiations to negotiate emission reduction targets for years beyond 2012 was inevitable for the international community to tackle climate change. However, at the COP meeting under the UNFCCC, most developing countries, as well as the United States, considered that it was too early to start such negotiations, and thus the COP agreed to start a two-year dialogue to exchange views on this matter. Reflecting the view of the developing countries and the United States, it was confirmed that the dialogue would "not open any negotiation leading to new commitments" (UNFCCC 2006). On the other hand, the situation was quite different at CMP1 under the Kyoto Protocol. Article 3.9 of the Kyoto Protocol states that "commitments for subsequent periods for Parties included in Annex I shall be established in amendments to Annex B to this Protocol, which shall be adopted in accordance with the provision of Article 21.7," and that "the CMP shall initiate the consideration of such commitments at least seven years before the end of the first commitment period referred to in paragraph 1 above." The end of the first commitment period had been set to be 2012, so it was inevitable that the CMP would initiate negotiations on this matter in 2005. That was how a new round of negotiations started, to discuss Annex I countries' emission reduction targets for the years beyond 2012.

The process was called the Ad hoc Working Group on Further Commitments for Annex I Parties under the Kyoto Protocol (AWG-KP).

Two years of negotiations under AWG-KP made little progress. Annex I countries that were parties to the Kyoto Protocol had no incentive to negotiate their own emission reduction targets as long as the United States and developing countries were beyond the scope of the discussions. Two years of dialogue under COP also made little progress because it was expected that new negotiations would not be opened. Meanwhile, many activities outside the UNFCCC arena helped countries, regions, and local governments to move forward.

The G8 process achieved progress in the climate change debate. When the United Kingdom hosted the Gleneagles Summit in 2005, the then Prime Minister Tony Blair believed that climate change should be one of the key issues on the G8 agenda. The meeting initiated a three-year process that aimed at the United States' full involvement in dialogue and agreement on a long-term global goal. The process was successful in raising the awareness of heads of states concerning climate change. The G20 process ran parallel to the G8 process, and has become increasingly influential, as it involved many newly developed economies.

The European Union (EU) initiated a regional emissions trading scheme in January 2005. Independent of the mechanisms set by the Kyoto Protocol, the EU reasoned that the use of a carbon market could be the most economically efficient instrument for achieving a certain level of targeted emission reductions.

The IPCC published its fourth assessment report (AR4) in early 2007. The IPCC, together with former US Vice President Al Gore, won a Nobel Peace Prize in the same year. These activities toward combating climate change attracted attention at the international level. By the end of 2007, many developing countries as well as the United States were ready to start negotiations on future international cooperation on climate change.

The COP13 and CMP3 were held in Bali in late 2007. Almost all countries supported starting negotiations, but each country held different ideas on what the agenda should be for the next round of negotiations. Industrialized countries emphasized that developing countries' mitigation actions should be negotiated. Developing countries considered it more important to see tangible progress in discussions on matters related to adaptation, finance, technology transfer, and capacity building, which were part of the responsibility of the industrialized countries under the UNFCCC. An agreement called the Bali Action Plan (BAP) was agreed on the final day of the COP13. The BAP aimed at agreeing on a comprehensive agreed outcome in COP15/CMP5 to be held in 2009. The countries were to negotiate on a wide variety of issues including a shared vision, mitigation actions for developed countries including setting emission reduction targets, nationally appropriate mitigation actions (NAMAs) for developing countries, reducing emissions from deforestation and forest degradation in developing countries (REDD), sectoral approaches, adaptation, finance, technology transfer, and capacity building. The new negotiation process following the BAP was called the Ad hoc Working Group on Long-term Cooperative Action under the Convention (AWG-LCA).

Countries experienced another two years of intense negotiations on two parallel tracks: AWG-LCA and AWG-KP. AWG-LCA's negotiating text comprised more than 200 pages due to the comprehensiveness of BAP's agenda. Another point of complexity was the relationship between the two tracks. The BAP called for "an agreed outcome" without defining the outcome's legal framework. It did not clearly state whether the outcome ought to be a new protocol, an amendment to the UNFCCC, or a set of COP decisions. On the other hand, the very existence of the AWG-KP process itself implied the continuation of the Kyoto Protocol and possible agreement on the second commitment period. In this case, what would the relationship be between the countries' commitments under the COP and the CMP? No one had an answer for this.

As the end of 2009 drew near, the Danish government, the host of the COP15/CMP5, called for a high-level segment at the heads-of-state level. The Danish government predicted that government officials could only take negotiations so far, and believed that the final part needed to be solved by ministers. The COP15/CMP5 was held in Copenhagen in December 2009. More than 40,000 people registered to participate in the meeting. In the final days of the Copenhagen meeting, 119 heads of states also participated in the meeting. Although the event attracted attention globally, the meeting failed to reach an agreement on a future climate institution. Both the AWG-LCA and AWG-KP were extended for another year to seek an agreement at COP16/CMP6. The only outcome was that the COP took note of the political agreement, which was called the Copenhagen Accord (UNFCCC 2010). The structure of the Copenhagen Accord followed that of the BAP.

The COP agreed that their shared vision for long-term cooperative action was to hold the increase in global temperature below 2 degrees Celsius and to achieve peaking of global GHG emissions as soon as possible.

As for mitigation, Annex I parties committed to implement quantified economy-wide emissions targets for 2020 and to submit these to the secretariat by the end of January 2010. Annex I Parties that were also Parties to the Kyoto Protocol were requested to further strengthen the emissions reductions initiated under the Protocol. Meanwhile, the transparency of actions on NAMAs by developing countries was discussed intensively. Industrialized countries wanted to see emission mitigation by developing countries. For emission mitigation to be observable, a process to ensure transparency became a core issue at the Copenhagen meeting. The process was called measurement, reporting, and verification (MRV). Some developing countries did not accept the MRV process, saying that such a process would impair national sovereignty. The Copenhagen Accord decided the non-Annex I Parties to the Convention would implement mitigation actions, and information on the actions was to be submitted to the secretariat by the end of January 2010. NAMAs taken by non-Annex I countries would be subject to their own domestic MRV. These countries would communicate information on the implementation of their actions through national communications, with provisions for international consultation and analysis (ICA). On the other hand, NAMAs that were internationally supported would be subject to international MRVs.

REDD, including REDD-plus (REDD, plus forest conservation and sustainable forest management), was one of the few issues that was supported by both the developed and developing countries. Developed countries were called on to mobilize financial resources. Further discussions were considered necessary in order to implement REDD so that the scheme would be able to accept credits for emission mitigation resulting from REDD activities.

The negotiations made great progress in the area of finance. A total of US$30 billion was announced as funding called "fast start finance" for the period 2010–2012. Long-term financing was expected to reach USD 100 billion a year by 2020 to address the needs of developing countries. A significant portion of such funding was expected to flow through a new funding mechanism called the Copenhagen Green Climate Fund.

In a way, the Copenhagen Accord provided answers to what had been listed in the BAP, but for procedural reasons the COP merely took note of it rather than adopting it. The Copenhagen Accord was drafted with the involvement of a limited number of countries, including the United States and China. The heads of state of many small developing countries were excluded from the drafting process, and information was not shared with all the countries. On the final day of the COP, several countries, such as Bolivia and Venezuela, strongly opposed the adoption of the Copenhagen Accord. They criticized the decision-making process at the Copenhagen meeting, which was said to have been exclusive and opaque, and thus illegitimate. For other observers, the UNFCCC process, which was blocked by a few countries, was perceived to be detrimental.

In 2010, negotiations under AWG-LCA and AWG-KP continued for another year. When countries gathered at Cancun in late November 2010 to attend COP16/CMP, they had no idea how the meeting would evolve. Many thought the final agreement could be reached at COP17/CMP7 in 2011. At the very least, the Cancun meeting was expected to anchor the context of the Copenhagen Accord in the COP/CMP documents. The negotiation texts under the two parallel tracks continued numerous parenthetical expressions describing various options.

Contrary to the low expectations before the meeting, the Cancun meeting was able to achieve what it had been expected to do. One difference from COP15/CMP5 was the features of the process. The Mexican government, led by Patricia Espinosa, President of COP16/CMP6, maneuvered the participants through the complicated negotiation process by emphasizing transparency, inclusiveness, and legitimacy. She regularly emphasized the fact that there was no "Mexican text" for the negotiations. All countries were updated regularly on the latest news concerning the status of negotiations. On the final day, the President's proposal was accepted without any revisions. Bolivia's opposition was taken note of. The whole set of COP/CMP decisions was called the Cancun Agreement. The context of the Cancun Agreement follows that of the Copenhagen Accord. By anchoring the political agreement as COP/CMP formal decisions, the Copenhagen Accord was finally part of the formal UNFCCC process. The Cancun Agreement laid out what required further negotiation by COP17/CMP7, to be held in Durban in 2011. Countries praised President Espinosa's skillful handling, saying that multilateralism had recovered.

A key feature of the Cancun Agreement under the AWG-LCA was that it readdressed what had been agreed to in the Copenhagen Accord. This was necessary because without this procedure the Copenhagen Accord would have been no more than a political agreement independent of the formal UNFCCC negotiation process. The main content was as follows (UNFCCC 2011).

- A shared vision for long-term cooperative action called for deep cuts in global GHGs with a view to hold the increase in global average temperature below 2 degrees Celsius above pre-industrial levels. It also noted the need to consider strengthening the long-term global goal, given the global average temperature rise of 1.5 degrees Celsius.
- With regard to adaptation, the COP agreed to establish the Cancun Adaptation Framework to invite developing countries to plan and implement adaptation actions. It also requested that the developed countries provide financial and technical support to the least developing countries that have difficulty in fulfilling the adaptation activities.
- With regard to mitigation actions by the developed countries, it was agreed that the quantified economy-wide emission reduction targets to be implemented by Annex I parties communicated by them under the Copenhagen Accord were to be described in a subsidiary body's informal document (an "INF" document), and that the COP would take note of it. This was a way to anchor the political pledge in the formal UNFCCC process. In addition, the developed countries were also urged to increase the ambition of their targets to a level consistent with the IPCC AR4.
- With regard to mitigation action by the developing countries, it was agreed that the NAMAs to be implemented by non-Annex I parties communicated by them under the Copenhagen Accord were to be described in an INF document of AWG-LCA, and that the COP would take note of it. The developing countries would submit their national communication in accordance with Article 12 of the UNFCCC every four years. They would also submit biennial update reports. Actions taken with domestic support would be measured, reported, and verified (MRVed) domestically in accordance with general guidelines to be developed under the Convention, and the results of this process would further be subject to ICA. Internationally supported actions would be MRVed domestically and would be subject to international MRVs in accordance with guidelines to be developed under the Convention.
- With regard to finance, it was reaffirmed that the fast start finance for the period 2010–2012 would reach US$30 billion, with a balanced allocation between adaptation and mitigation. As for the long-term financing, it was agreed to aim for a goal of jointly mobilizing USD 100 billion per year by 2020, on condition that the funds may come from a wide variety of sources,

both public and private. The Green Climate Fund was established, and was to be designated as an operating entity of the financial mechanism of the UNFCCC; its board would comprise 24 members with an equal number of members from developing and developed countries. The Transitional Committee to the Green Climate Fund was to have 40 members, with 15 members from developed countries, and 25 members from developing countries, with a designated number from each region.

- With regard to technology, the Climate Technology Center and Network was established. Its details were to be further negotiated prior to COP17.

Despite everything that was achieved in the Cancun Agreement, however, it is also true that most of the core issues were left untouched in Cancun. The level of emission reduction pledges by the developed countries is still far less than what is considered to be necessary to achieve the long-term target of 2 degrees. The United States, the largest emitter among the developed countries, is unwilling to accept an emission reduction target as being internationally legally binding in nature. Many developed countries do not wish to have their respective emission reduction targets made legally binding unless the United States accepts the same type of obligations as other developed countries. In the United States, legislation on climate change mitigation policies has not gained public support.

Developing countries still confront the developed countries with their historical responsibilities. China is not happy with the MRV process laid out by the UNFCCC process. Discussions on financial mechanisms have arrived at the establishment of a new fund, but pledges of developed countries have not reached the level of the amount that was agreed in the Copenhagen Accord.

Negotiations have not yet touched upon the legal nature of the agreement. Japan and Russia are unhappy with the continuation of the current Kyoto Protocol, as the GHG emissions of the United States and emerging economies are not regulated by the Protocol. They assert that a new protocol, in which all countries participate, should take over from the current Kyoto Protocol to integrate the two tracks. Many developing countries strongly oppose this idea, as the Kyoto Protocol is currently the only multilateral agreement that lays out emission reduction targets for developed countries, and thus maintain that it should not be abolished so easily. All of the core issues remain hurdles that will be difficult to overcome at COP17/CMP7.

Climate change mitigation in Asia

According to the Fourth Assessment Report (AR4) published in 2007, Asia is likely to be affected by climate change in various ways (IPCC 2007). First, because many countries in Asia have long coastlines, they are likely to be affected by rising sea levels or heavy rainfall, and coasts are already eroding. Crop yields of rice, grain, and corn are likely to be affected by changes in precipitation patterns and temperature. Some regions could benefit from climate change, while many other regions will see lower crop yields due to less rainfall and higher temperatures. Countries in Asia have already started to be affected by the impact of climate change, and it is likely that the degree of adverse impact will increase in the near future.

GHG emissions from countries in Asia are growing rapidly. Figure 2.1 shows the share of CO_2 emissions from fuel combustion by countries in 2007. Emissions from China, India, Japan, and Korea add up to around 31.6 percent of global emissions. Any regional policies that may reduce emissions in Asia could make a difference to one-third of global emissions.

At the same time, it should be noted that graphs using emission per capita deliver a message from another angle. Figure 2.2 shows the size of the population in each region and the region's GHG emissions per capita in terms of CO_2 equivalents. Note that this figure was taken from the AR4 of the IPCC published in 2007, and thus the data are for the year 2004, which makes the data inconsistent with Figure 2.1. The purpose of this figure is to illustrate the wide gaps in the levels of emissions per capita between regions around the globe. The size of each rectangle shows the scale of emissions from the region. It is true that the size of the US and Canada is almost the same as the size as Non-Annex I East Asia, but the latter's per capita emissions are much lower than the former's. The same can be said for South Asia as well. From Figure 2.2, we see that many Asian countries are still less developed, and thus it should be countries in the developed world that should start reducing their GHG emissions earlier than others.

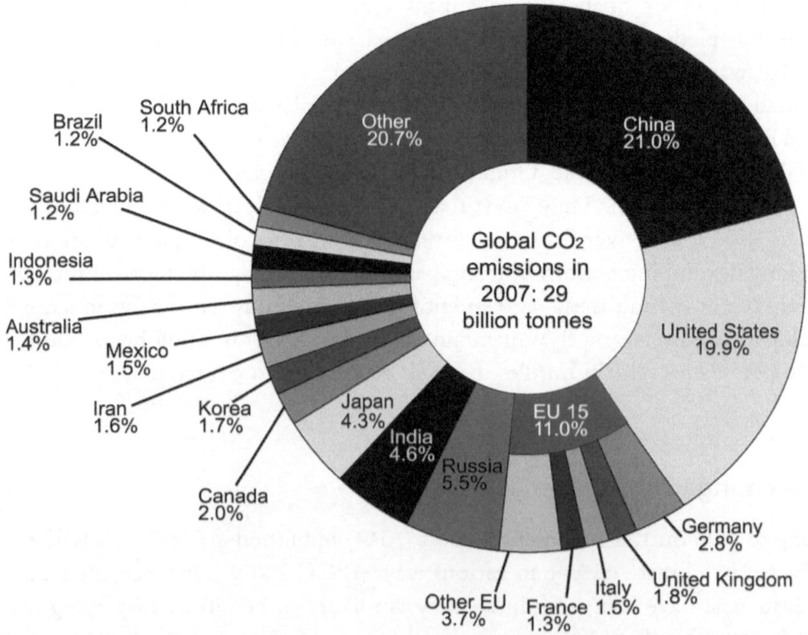

FIGURE 2.1 CO_2 emissions from fuel combustion by countries in 2007 (source: data from IEA).

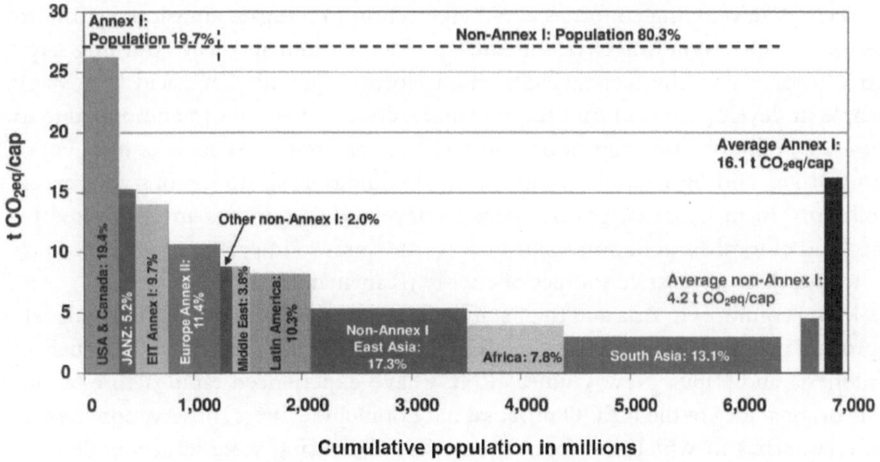

FIGURE 2.2 Year 2004 distribution of regional per capita GHG emissions (all Kyoto gases, including those from land use) over the population of different country groupings (source: IPCC 2007).

Note
The percentages in the bars indicate a region's share in global GHG emissions (IPCC 2007).

Based on what is shown in Figure 2.2, it is fair to say that Asian developing countries should be allowed a certain amount of further emission growth. On the other hand, it is also true that the absolute level of emissions from these countries occupies a large share of global emissions. Neither the UNFCCC nor the Kyoto Protocol set legally binding emission reduction targets for developing countries, and many developing countries have already started to take action voluntarily. There are several reasons for developing countries to take action to mitigate emissions without making commitments internationally.

First, the rapidly growing economies, where the demand for energy has been growing at the same time, face an energy supply shortage. China, for example, used to be an oil-exporting country, but it became an oil-importing country in 1993 (Shaffer 2009), and the amount it imports has been increasing continuously. Considering the high international price of oil, it is economically wise for China to improve its energy efficiency. In many cases, improving energy efficiency is the most cost-effective way to reduce CO_2 emissions. This is also the reason why developing countries are interested in technology transfer for improving energy efficiency.

Similarly, improving energy efficiency or investing in alternative sources of energy are ways to maintain a country's energy security. India, for example, has been investing in solar energy: although solar power and many other forms of renewable energy are much more expensive than traditional sources of energy, such as coal and fuel woods, it is one means of obtaining purely domestic energy that will not run out. Similar types of investment could be made for recovering methane (CH_4) from landfill sites.

Second, developing countries could take action to mitigate emissions actions to deal with other types of environmental problems simultaneously. Burning large amounts of coal in low-tech facilities leads to local air pollution and acid rain. Many people in developing countries suffer from health issues such as pneumonia due to local air pollution. Burning noncommercial energy resources such as fuel wood, dung cake, and agricultural waste inside buildings leads to serious indoor air pollution. In many cases, governments in developing countries are motivated by this kind of local air pollution to start taking measures to improve energy efficiency or to switch to alternative sources of energy (Ramanathan 2003).

Many countries in Asia and the Pacific region are also interested in recent developments in the debates concerning REDD and REDD-plus. Countries such as Indonesia and Papua New Guinea (PNG) have experienced rapid deforestation. The original idea of the REDD proposal did not include forest conservation: rather, only countries in which the forest areas were decreasing were targets under this scheme. On the other hand, the REDD scheme has paved the way for countries such as China and India to be involved in emission mitigation policies, while introducing a new option for protecting their forests.

Third, active participation in international negotiations on environmental problems can be considered to be important from a diplomatic point of view. A country may take positive action on climate change if such action could improve the diplomatic standing of that country. Negotiating positions formulated by ministries of foreign affairs can be relatively easily affected by diplomatic circumstances (Yu 2008). For many developing countries, climate change negotiations may involve negotiations on funding rather than on GHG emissions. Financial assistance and technology transfer can be negotiated by officials in charge of foreign affairs or development, rather than the environment. For some major emerging economies, climate change negotiations are one way to gain political power at the international level (Zhang 2003).

Fourth, developing countries may wish to take action to tackle climate change once they know that such action can receive financial support from international financial mechanisms. Several funding mechanisms are established under the UNFCCC and the Kyoto Protocol. There is also a mechanism called the Clean Development Mechanism (CDM), which channels financial support and technology transfer to emission mitigation projects in developing countries by raising certified emission reductions (CERs), which are a kind of carbon credit. Up to now, the CDM scheme has made great contributions to mobilizing assistance to emission mitigation policies in developing countries. As of January 2011, 2,777 projects had been registered. The problem with the CDM lies in the distribution of projects among developing countries, as 42.7 percent of the registered projects were hosted in China, and 22.1 percent in India (UNFCCC Secretariat 2011).

Last but not least, developing countries can introduce emission mitigation policies to support domestic industries whose products could sell well in the world of low carbon development (LCD). Manufacturing photovoltaic solar panels and wind turbines could be regarded as a great opportunity for new business. New industries could be supported by their government to gain international competitiveness.

For the various reasons mentioned above, many developing countries assert that they have already started to voluntarily take action to mitigate emissions, denying criticism from developed countries that they are not making international commitments under the UNFCCC. The conflict actually became the central issue at COP15, when developed countries, especially the United States, demanded a process for transparency concerning mitigation actions in developing countries. Many countries in Asia had self-pledged emission reduction targets around the time of COP15. Those targets are considered to be voluntary targets and not international commitments, but they are submitted to the UNFCCC secretariat under the provision of responding to what was requested by the Copenhagen Accord. Table 2.2 shows some of the emission mitigation targets and actions submitted by countries in Asia and the Pacific region after COP15.

As can be seen in Table 2.2, many developing countries have adopted efficiency improvement targets, such as emissions per gross domestic product (GDP). This is a way to get around the issue of the inter-relationship between CO_2 emissions and economic growth. The aim of the UNFCCC is not to stall the economic growth of developing countries, but rather to enhance economic development while keeping GHG emissions as low as possible. Moreover, it is often the case that a

TABLE 2.2 Emission targets of countries in Asia and the Pacific region

Country	*Voluntary target*
China	Reduce CO_2 emissions per GDP by 40–45% by 2020 from 2005 levels.
India	Reduce CO_2 emissions per GDP by 20–25% by 2020 from 2005 levels.
Japan	Reduce GHG emissions by 25% by 2020 from 1990 levels, conditional upon major emitting countries participating in international mitigation agreements.
Korea	Reduce CO_2 emissions by 30% by 2020 compared with business-as-usual conditions.
Indonesia	Reduce CO_2 emissions by 26% by 2020 compared with business-as-usual conditions. With international assistance, the target will be changed to 41%.
Singapore	Reduce CO_2 emissions by 16% by 2020 compared with business-as-usual conditions.
Papua New Guinea	Reduce CO_2 emissions by 50% by 2025 (base year not defined).

Source: author.

country's absolute emissions fluctuate every year due to economic conditions. Setting a target for emissions per GDP will enable governments of developing countries to concentrate on efficiency.

One serious issue that will occur in using emissions per GDP as national targets is the difficulty it creates in the MRV process. In the event that a GHG emission target is set at an absolute level, MRV of GHG emissions alone would suffice. In the event that a target is set for emissions per GDP, MRV of not only the emissions but also of GDP would be required. The value of GDP changes depending on what type of GDP is to be used: there are both nominal and real GDPs. It would become even more complicated if the GDP were to be translated into another currency, such as US dollars, in order to be comparable with those of other countries. It usually takes one or two years for countries to be able to announce official GDP figures, and there are no good ideas on how to deal with this time lag.

Another type of emissions target set by many developing countries is a target based on the level of deviation between actual emissions and business-as-usual emissions. For example, Singapore is to reduce its CO_2 emissions by 16 percent compared with business-as-usual emissions. We have yet to figure out, however, how to calculate emissions under a business-as-usual scenario: it cannot be measured because it is a virtual figure.

Although the above-mentioned issues remain, we may still be able to say that the targets set by developing countries are a great step forward. Developing countries have announced their actions, and now they are ready to discuss procedural matters to secure transparency with the data on policies and measures and their effectiveness in reducing GHG emissions.

Adaptation is also very important for countries in Asia and the Pacific region. Many small island states in the region will be affected by rising sea levels. Some low-lying countries such as Bangladesh will be affected as well. Nepal worries about the melting of glaciers in the Himalayas. In many cases, adaptation policies are beneficial for these countries even without the impact of climate change. Being fully prepared for extreme weather patterns will protect the community even if the extreme weather is not caused by climate change.

Ways to achieve low carbon development in Asia and the Pacific region

Although major international institutions to deal with climate change have been developed at the multilateral level under the auspices of the United Nations, cooperation at the regional level also has its merits. First, each region has regional circumstances that are not shared by other regions. In the case of Asia and the Pacific region, the region is different from other parts of the globe because of its rapid economic growth, as well as its rapid urbanization.

The relationship between CO_2 emissions and GDP had been laid out by developing countries by setting a target for the ratio between the two, but in reality it is not that simple. Figure 2.3 shows the correlation between climate change and

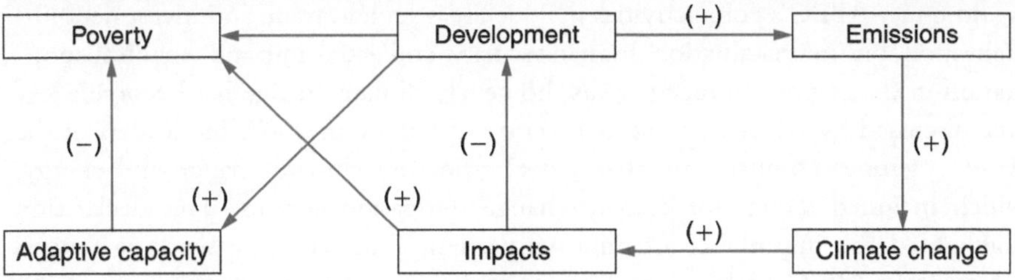

FIGURE 2.3 Correlation between climate change and poverty (source: Sari 2008: 7).

poverty. Economic development may reduce poverty, but the quality of life of the poor may be aggravated by the increase in the adverse impacts of climate change, such as floods and droughts. Even if developing countries succeed in achieving their emission per GDP targets, the rapid growth of CO_2 emissions at the absolute level may accelerate climate change, leading to more adverse impacts of climate change that could harm their own development.

These are the main reasons why it is now becoming more important to seek a comprehensive approach to the problem. The aim is not to reach a certain target set under the UNFCCC, but to achieve low carbon development. Although there is no clear definition of what low carbon development is, its goal is to reduce GHG emissions to mitigate climate change, while enhancing a better and happier way of life through achieving qualitative development. This could be achieved by, for example, constructing a town where the urban design makes it possible to do away with the use of private automobiles. Shopping malls, schools, and offices could be built within walking distance so that local residents do not feel inconvenienced. If a good security system is introduced, such a town could be a much safer place to live than a situation in which people live far apart from each other. Renewable energy can be introduced in these urban areas.

As multilateral negotiations on a post-2012 institution under the UNFCCC and the Kyoto Protocol have been making little progress, regional cooperation in Asia and the Pacific region is becoming more important. The Asian Development Bank (ADB) has prepared a financial scheme to help countries in Asia to implement climate change policies. The ADB is an international development finance institution. Its main mission is to help its developing country members to reduce poverty and improve the quality of life of their people. In recent years, however, much of the ADB's support has shifted to tackling climate change or other environmental problems while reducing poverty and improving the people's quality of life. Developing infrastructure for public transportation, or building dams for hydropower plants, could raise the quality of life of the people while at the same time reducing CO_2 emissions.

The Asia Pacific Economic Cooperation (APEC) forum was originally intended to facilitate economic cooperation among 21 nations and regions in the Asia-Pacific area. The scale of APEC's member economies accounts for around 50 percent of the world's GDP, and about 40 percent both of the world's population and of total world trade; APEC's core activities include areas such as trade and investment liberalization, business facilitation, human security, and economic and technical cooperation in the region. In recent years, however, climate change has become a key area discussed by APEC, and at its meeting in Sidney in 2007, the leaders of the APEC member countries adopted a declaration on climate change and energy, which included targets for climate change mitigation actions. This declaration highlighted the importance of improving energy efficiency by working toward achieving an APEC-wide regional aspirational goal of reducing energy intensity by at least 25 percent by 2030 (with 2005 as the base year), and by increasing forest cover in the region by at least 20 million hectares of all types of forests by 2020.

In addition to these kinds of regional cooperation, many activities are implemented via bilateral cooperation. Japan, for example, has been assisting climate mitigation policies in many countries in the Asia-Pacific region by means of forest conservation, investments in solar power generation and efficient combined-cycle thermal power generation, and training and capacity building of local people.

Conclusion

Asia and the Pacific region will continue to be crucial world regions in terms of both climate change mitigation and adaptation. As multilateral negotiations on climate change are likely to face difficulties in achieving a comprehensive multilateral agreement for the post-2012 period, regional activity will become more and more important in the near future, at least in the short term. Asia and the Pacific region as a whole need to seek ways to achieve a low carbon society (LCS). How can it be done?

First, instead of starting with the climate system or the distribution and use of carbon at the global level, we can focus on individuals at the local level: their everyday livelihoods are where we can find primary solutions. What are their energy use patterns? What are their means of transportation? Do the people in a particular place face certain kinds of local environmental degradation or local pollution? How are they affected? How do people get information on climate change and on the ways and means of mitigation? What are the major industries and businesses that bring about economic growth in the society? How can these industries and businesses be de-carbonized without harming their level of economic development? What are the features of public transportation in the society? How can it be better used by local people? It is often the case that a high level of security in a society is a precondition for people to use public transportation. This means that safety, or a peaceful society, is one of the conditions required for LCS to be achievable (Conca 1996). What can be said for the society's waste management and recycling systems? Reduction of waste, together with the use of energy contained in the waste, is a way to link local waste management and climate change mitigation. How have policies on conservation of biodiversity been introduced in the society? Protection of forests is also a way to meet both climate and biodiversity goals.

Second, we need to focus on decision making at a national level. What is the country's energy policy? How does its government view energy security? In countries with abundant coal resources, the sustainable use of coal, rather than investments in renewable energy technology, will be considered to be a way to achieve energy security. In such cases, carbon capture and storage (CCS) might be a key technology necessary to achieve LCS. What is the government's view of the foreign policy dimension of the climate change problem? The UNFCCC negotiations have been dealing with financial mechanisms and technology transfer for a long time, without making much progress. What would be necessary for developing countries to be able to benefit from international assistance in terms of emission mitigation policies?

Last but not least, it would be worthwhile to analyze what could be achieved at a regional level. For example, an electricity supply could be provided by building an up-to-date smart grid throughout a region with a high population density. Local towns with low population densities could be supplied with electricity via micro grids using local energy. The same is true for transportation: a railroad network could be built across national borders so as not to increase the use of air transportation. Another example is waste management and recycling, which could be most effective if handled at a regional level rather than a national level. Waste in one country could become a resource in another country.

In the foreseeable near future, developing countries will not commit to any emission mitigation targets as an international commitment under the UNFCCC. The focus of the negotiations will be on the process of securing the transparency of NAMAs. On the other hand, what is needed for climate change mitigation is to accelerate the use of NAMAs, which requires incentives outside the rules set by the UNFCCC. This chapter was intended to review the whole range of activities outside the UNFCCC that are aimed at achieving LCS in developing countries. In any case, the reduction of GHGs needs to coexist with sustainable development, and this needs to be done without losing any more time.

References

Agarwal, A. (2002) "A Southern Perspective on Curbing Global Climate Change," in Schneider, S., Rosencranz, A., and Niles, J. (Eds.), *Climate Change Policy: A Survey*. Washington, DC: Island Press, 375–91.

Conca, K. (1996) "Peace, Justice, and Sustainability," in Pirages, D. (Ed.), *Building Sustainable Societies*. Armonk, NY: M.E. Sharpe, 17–31.

Grubb, M., Vrolijk, C., and Brack, D. (1999) *The Kyoto Protocol: A Guide and Assessment.* London: The Royal Institute of International Affairs, Earthscan.

IPCC (Intergovernmental Panel on Climate Change) (2007) *Climate Change 2007: Impacts, Adaptation and Vulnerability.* Contribution of Working Group II to the Fourth Assessment Report of the Intergovernmental Panel on Climate Change. Cambridge: Cambridge University Press, 976pp.

Mintzer, I.M. and Leonard, J.A. (Eds.) (1994) *Negotiating Climate Change: The Inside Story of the Rio Convention.* Cambridge: Cambridge University Press.

MOEJ (Ministry of the Environmental, Government of Japan) (2011) Online information. Online, available at: www.env.go.jp/earth/ondanka/knowledge.html (accessed August 25, 2011).

Oberthür, S. and Ott, H. (1999) *The Kyoto Protocol: International Climate Policy for the 21st Century*. Berlin: Springer.

Ramanathan, R. (2003) "Evaluating Greenhouse Gas Control Strategies Using Multicriteria Approaches," in Toman, M., Chakravorty, U., and Gupta, A. (Eds.), *India and Global Climate Change: Perspectives on Economics and Policy from a Developing Country*. Washington, DC: Resources for the Future, 195–216.

Sari, A.P. (2008) "Introduction: Climate Change and Sustainable Development in Asia," in Kameyama, Y., Sari, A., Soejachmoen, M.H., and Kanie, N. (Eds.), *Climate Change in Asia: Perspectives on the Future Climate Regime*. Tokyo: United Nations University Press, 3–17.

Shaffer, B. (2009) *Energy Politics*. Philadelphia: University of Pennsylvania Press, 143–8.

UNFCCC (United Nations Framework Convention on Climate Change) (1995) "Decision 1/CP.1," in *Report of the Conference of the Parties on its First Session: 1995 Document FCCC/CP/1995/7/Add.1*, Berlin, Germany, March 28 to April 7, 1995. New York: United Nations.

UNFCCC (United Nations Framework Convention on Climate Change) (2002) *Report of the Conference of the Parties on its Seventh Session: Document FCCC/CP/2001/13/Add.1&Add. 2*, Marrakesh, Morocco, October 29 to November 10, 2001. New York: United Nations.

UNFCCC (United Nations Framework Convention on Climate Change) (2006) "Decision 1/CP.11," in *Report of the Conference of the Parties on its Eleventh Session: Document FCCC/CP/2005/5/Add.1*, Montreal, Canada, November 28 to December 10, 2005. New York: United Nations.

UNFCCC (United Nations Framework Convention on Climate Change) (2010) "Decision 2/CP.15," in *Report of the Conference of the Parties on its Fifteenth Session: Document FCCC/CP/2009/11/Add.1*, Copenhagen, Denmark, December 7–19, 2009. New York: United Nations.

UNFCCC (United Nations Framework Convention on Climate Change) (2011) "Decision 1/CP.16," in *Report of the Conference of the Parties on its Sixteenth Session: Document FCCC/CP/2010/7/Add.1*, Cancun, Mexico, November 29 to December 10, 2010. New York: United Nations.

UNFCCC Secretariat (2011) CDM in Numbers. Online, available at: http://cdm.unfccc.int/Statistics/Registration/NumOfRegisteredProjByHostPartiesPieChart.html (accessed August 25, 2011).

Yu, Hongyuan (2008) *Global Warming and China's Environmental Diplomacy*. New York: Nova Science Publishers.

Zhang, Z. (2003) "The Forces behind China's Climate Change Policy: Interests, Sovereignty, and Prestige," in P.G. Harris (Ed.), *Global Warming and East Asia*. London: Routledge, 66–85.

3

CURRENT STATUS AND CRITICAL ISSUES OF THE CDM

Kazuhito Yamada and Mariko Fujimori

What is the CDM?

The Clean Development Mechanism (CDM) is a project-based mechanism stipulated in the provisions of Article 12 of the Kyoto Protocol. The CDM is a scheme to realize greenhouse gas (GHG) emission reduction through cooperation between Annex I Parties to the United Nations Framework Convention on Climate Change (UNFCCC). The Annex I Parties are committed to certain GHG emission reduction targets under the Kyoto Protocol (so-called developed country), and non-Annex I Parties do not have any commitments to reduce GHG emissions (so-called developing country). The purposes of the CDM are to assist in accomplishing the GHG reduction targets of Annex I Parties (so-called investing countries), as well as to contribute to the sustainable development of non-Annex I Parties (so-called host countries). Under the CDM, basically project participants in Annex I Parties implement GHG emission reduction projects in cooperation with project participants in non-Annex I Parties within the territories of the latter Parties. Annex I Parties are eligible for all or part of the credits (certified emission reductions: CERs) resulting from the projects. Non-Annex I Parties will benefit (economically, socially, environmentally, and technologically) from CDM projects (Figure 3.1).[1]

Under the CDM, since the CERs are generated in host countries that do not have any commitments to reduce GHG emissions, the total amount of permitted emissions in the Annex I Parties is increased. Lower actual reductions than issued CERs would result in increased GHG around the world. Therefore, the CDM requires projects to follow strict procedures in setting baseline scenarios specified by the CDM Executive Board (CDM EB) to ensure that the GHG emission reductions are additional by comparison and that CERs are not overestimated. The procedures include validation and verification by third-party emission reductions

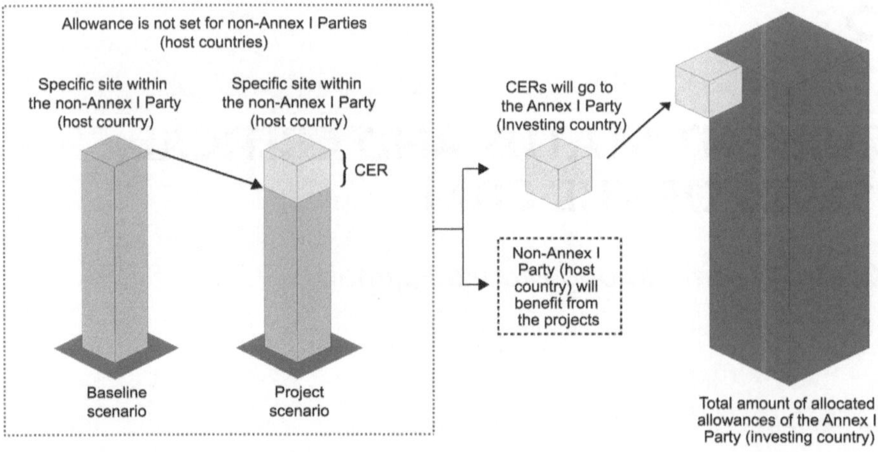

FIGURE 3.1 Outline of the CDM (source: MOEJ, 2010).

assessment using designated operational entities (DOEs), and a final approval of project registration with the issuance of CERs by the CDM EB. Article 12 of the Kyoto Protocol stipulates that emission reductions of each CDM project shall be certified by the DOE on the basis of:

a. voluntary participation approved by each Party involved;
b. real, measurable, and long-term benefits related to the mitigation of climate change; and
c. reductions in emissions that are additional to any that would occur in the absence of the certified project activity.

The CDM EB has developed a series of modalities and procedures to emphasize the importance of these three requirements for implementing CDM projects.

Procedures and key concepts of the CDM

The CDM project cycle is as follows. Project participants wishing to realize CDM projects must follow each procedure according to the instruction materials prepared by the CDM EB.

Developing and designing the CDM project

The CDM projects can be classified as illustrated in Table 3.1. After confirming project type, scale, and activities, project participants need to draft a project design document (PDD) describing the details of the project according to the templates of PDD developed by CDM EB as well as baseline application and monitoring methodology.

TABLE 3.1 Types, scales, and activities of the CDM projects

Type	*Scale*	*Activities*
Emission reduction type CDM	Large-scale CDM	CDM project activities which reduce GHG emissions from fossil fuel combustion, fossil fuel consumption, mining, and fugitive emission such as HFC23, N_2O, and CH_4 in host countries.
	Small-scale CDM (SSC)	Type I: renewable energy projects with a maximum output capacity of 15 MW (or an appropriate equivalent).
		Type II: energy efficiency improvement projects which reduce energy consumption, on the supply and/or demand side, with a maximum output of 60 GWh per year (or an appropriate equivalent).
		Type III: other projects such as transport and waste management that result in emission reductions of less than or equal to 60 kt CO_2 equivalent annually.
Removal type CDM/Afforestation and Reforestation (A/R) CDM	Large-scale A/R CDM	CDM project activities to result in GHG removals by sinks through afforestation and reforestation including mangrove planting in host countries.
	Small-scale A/R CDM (SSC A/R)	Those that are expected to result in net anthropogenic GHG removals by sinks of less than 16 kt CO_2 per year and are developed or implemented by low-income communities and individuals as determined by the host Party.

Source: CMP/2006/10/Add. 1, page 8, para. 28, CMP/2007/9/Add. 1, page 26.

Concerning the application of the baseline and monitoring methodology, project participants who want to apply an approved methodology (AM) to their project activity must check the applicability conditions of AMs and approved consolidated methodologies (ACMs). They may propose new baseline and monitoring methodologies, if they cannot find a suitable methodology for their project.[2]

Getting approval letters from the designated national authorities

CDM project candidates are required to obtain approval of all the Parties involved before they can be officially validated and registered as CDM projects. The Parties involved are usually both the host country and the investing country/countries. The approval of the host country shall be the written approval of voluntary participation which is issued from the designated national authorities (DNA), including confirmation by the host country that the project assists it in achieving sustainable development.[3]

It is possible to register a CDM project in which there are no investing country/countries involved. Such project activities are called "unilateral CDM" projects.

This means that approval by investing country/countries would not be necessary to register a CDM project. Investing country/countries can acquire CERs from a unilateral project activity after submitting a letter of approval regarding the project activity to the CDM EB.

The key concept at this stage is that "the project can assist in achieving sustainable development." The national procedures and process for the host country's approval for CDM projects are decided by each country. Those procedures could contain sustainable development criteria for the Party. The decision of the criteria is a prerogative right of host country. Therefore, UNFCCC and CDM EB do nothing, including recommendation or guidance about the criteria.

Implementing validation and registration of the CDM project

Project participants, in tandem with getting approval letters from the DNA, must implement validation of the PDD according to the initiative of the DOE. Validation is the process of independent evaluation of an appropriateness of a proposed project by a DOE against the requirements of the CDM on the basis of the PDD. Project participants should select and assign one appropriate DOE and make a contract to undertake validation. The first step of the validation is for project participants to submit a draft PDD to a DOE, which reviews the PDD and opens it for public input by disclosing a UNFCCC website. In the review process, the DOE draws the attention of the project participants to significant points in the PDD that need to be clarified and improved through "corrective action requests" and "clarifications" so that project participants can improve and finalize their PDD. The DOE finally determines whether the proposed project should be validated or invalid and then informs project participants.

Registration is the formal acceptance by the CDM EB of a project validated as a CDM project. The DOE submits to the CDM EB a request for registration in the form of a validation report including the PDD, the written approvals of both countries, and an explanation of how it has taken due account of comments received. The proposed project will be registered by the CDM EB on the day following a period of request for review; 28 days after the date of publication of the request for registration by the secretariat, unless a country involved in the project or at least three members of the CDM EB request a review of the proposed CDM project.

The key concepts at this stage are "baseline scenario" and "additionality." The baseline is defined as "the scenario that reasonably represents the anthropogenic emissions by sources of GHGs that would occur in the absence of the proposed project."[4] The difference between baseline GHG emissions and GHG emissions after implementing the CDM project (project GHG emissions) is the emission reductions (Figure 3.2).

The concept of "additionality" is closely related to that of "baseline," and has to be paid particular attention to when setting up a baseline scenario. But there is no official terminology of "additionality" in the Kyoto Protocol. It is considered that "A CDM project activity is additional if anthropogenic emissions of GHGs by

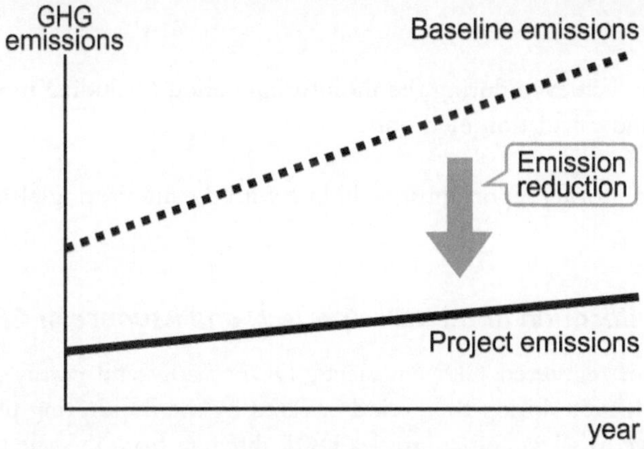

FIGURE 3.2 Concepts of emission reduction (source: MOEJ, 2010).

sources are reduced below those that would have occurred in the absence of the registered CDM project activity."[5]

Project participants must submit a written explanation of how and why their project is additional and therefore not the baseline scenario according to the instructions of the selected baseline methodology. The CDM EB provides "The tool for the demonstration and assessment of additionality" as a general framework for demonstrating and assessing additionality. This tool is to be applicable to a wide range of project types. Project participants can use this tool to assess and determine additionality but it does not replace the need for the baseline methodology to justify the selection and determination of the most plausible baseline scenario alternatives.

Monitoring the CDM project

The first and the most important step that project participants have to take to ensure issuance of CERs as planned in the PDD, is to properly implement the monitoring of the CDM project according to the monitoring plan stated in the PDD, since it is a condition for verification, certification, and the issuance of CERs. Project participants have to prepare a monitoring report for the certain period they would like to have CERs verified/issued. The length of the period can vary according to their interests (verified/issued periods of CERs range from one month to seven years).

"Guidelines for completing the monitoring report form" should contain information on the following items:

a. status of the implementation of the project activity;
b. actual operation of the project activity;
c. approved monitoring plan applied to the project activity;
d. monitoring procedures;
e. baseline emissions;

f. project emissions;
g. leakage emissions;
h. emission reductions achieved during the monitoring period (including monitored parameters and calculation methods).

The contents of the monitoring report must be in line with the monitoring plan of the registered PDD.

Verification and certification of the CDM project, and issuance of CERs

For getting CERs from registered CDM projects, DOEs verify and certify the emission reductions achieved during the period covered by the monitoring plan. Project participants have to select and contract a DOE different from the one that implemented validation of the corresponding project to implement verification and certification. In the case of small-scale CDM (SSC) projects, project participants can select and contract the same DOE to implement validation, verification, and certification. The basic documents for verification are in line with the above-mentioned guideline for monitoring.

Following verification, the DOE certifies in writing that the project achieved the GHG emission reductions as verified. It has to inform the project participants, Parties involved, and the CDM EB of its certification decision in writing immediately upon completion of the certification process, and make the certification report publicly available.[6]

A request for issuance of CERs is included in the certification report submitted to the CDM EB by the DOE. The DOE must submit the required documents listed in the completeness checklist for request for issuance on the UNFCCC CDM procedures. The request for issuance is finalized within 28 days after the date of publication of the request for issuance, unless a country involved in the project or at least three members of the CDM EB request a review of the proposed issuance of CERs. When the request for issuance becomes final or the CDM EB decides to approve the issuance as a result of a review, the CDM EB will instruct the CDM registry administrator to issue the specified quantity of CERs into the pending account of the CDM EB in the CDM registry.

Current status of the CDM projects

In order to promote and manage the CDM, CDM EB meetings have been held 58 times between 2001 and 2010. The Methodology Panel and the Small Scale Working Group have met 47 and 28 times respectively to develop methodologies and tools. As a result of these huge efforts, more than 100 methodologies and tools have been approved, and the institutional infrastructures to implement the CDM projects such as rules and references have gradually been established for potential project participants. Enjoying the benefit of these efforts, more than 2,800 CDM projects have been registered and almost 550 million CERs have been issued.

For a better understanding of the current status of the CDM projects, the authors have analyzed more than 2,500 registered CDM projects, and investigated the projects with CER issuance, based on the information from the UNFCCC website.

First, we analyzed the current status of the CDM projects focused on host countries. Table 3.2 shows that China has more than 40 percent of the registered CDM projects, followed by India, Brazil, and Mexico. China also has 56 percent of the CERs estimated to generate from registered CDM projects by 2012, followed by India, Brazil, and South Korea. Consequently, China has 56 percent of the issued CERs of registered CDM projects to date, followed by India, South Korea, and Brazil.

Second, we focused on the project types. Table 3.3 shows that renewable power projects such as hydropower (30 percent) and wind power (19 percent) dominated, followed by biogas, biomass, and CH_4 recovery in terms of registered project number. On the other hand, the second row of Table 3.4 shows that the non-CO_2 projects such as HFC23 destruction (25 percent) and N_2O destruction (13 percent) dominated, followed by renewable power and CH_4 recovery in terms of the CERs' estimate of generation from registered CDM projects by 2012. And the third row of Table 3.4 shows that the CERs issued by non-CO_2 projects such as HFC23 destruction (51 percent) and N_2O destruction (24 percent) accounted for three-fourths of the issued CERs of registered CDM projects.

Last, we calculated the estimated CERs per project types based on the data of registered CDM projects and the CERs estimated to generate from them by 2012. As shown in Table 3.5, the non-CO_2 project types of HFC23/N_2O destructions and CH_4 recovery/use have very high values of the estimated CERs per project.

TABLE 3.2 Share of registered CDM projects categorized by host country (%)

Host country	*Number of projects*	*Estimated CERs*	*Issued CERs*
China	41	56	53
India	22	14	17
Brazil	7	7	9
Mexico	5	3	2
Malaysia	3	1	1
Indonesia	2	1	<1
South Korea	2	5	13
Philippines	2	<1	<1
Thailand	2	<1	<1
Chile	2	2	1
Argentina	2	1	1
Egypt	<1	<1	1
Vietnam	<1	<1	1
Others	14	12	4

Source: UNFCCC homepage (as of December 2, 2010).

TABLE 3.3 Share of registered CDM projects categorized by project type (%)

Hydropower	30
Wind	19
Biogas	12
Biomass	11
Methane recovery and use	7
Waste heat and gas	7
Energy efficiency	3
N_2O	2
Fuel switch	2
Methane avoid	2
Others	5

Source: UNFCCC homepage (as of December 2, 2010).

TABLE 3.4 Share of CERs of registered CDM projects categorized by project type (%)

	Estimated	*Issued*
HFC23 destruction	25	51
Hydropower	14	5
N_2O destruction	13	24
Methane recovery and use	11	3
Wind	10	5
Waste heat and gas	8	5
Fuel switch	6	2
Biomass	5	3
Biogas	3	1
Others	5	1

Source: UNFCCC homepage (as of December 2, 2010).

TABLE 3.5 Estimated CERs per project types (unit 1,000 CERs by 2010)

HFC23 destruction	24,228
N_2O destruction	4,024
CH_4 recovery and use	1,231
Wind power	397
Hydropower	368
Biomass	339
Biogas	215

Source: author.

Critical issues of the CDM

Although the CDM seems to be successful because there have been more than 2,800 registered projects, the analysis of registered CDM project activities found that the projects are concentrated in "super-developing" countries such as China, India, and Brazil. The GHG emission reductions from these projects result primarily from non-CO_2 type project activities such as HFC23 destruction and N_2O destruction. The CERs issued by the non-CO_2 type projects account for three-fourths of the total CER issued until now.

Yamada (2008) showed that the concentration of CDM projects in super-developing countries was probably explained by Annex I countries' perceived advantages in those countries, such as the number and quality of investment choices and other factors like the availability of human resources, etc. Yamada (2008) also explained that the predominance of the non-CO_2 type projects was likely due to their higher profitability (emissions credits) because of their higher global warming potential (GWP), and that the ultimate potential for GHG emission reductions from these non-CO_2 type projects, however, was likely to be comparatively limited since non-CO_2 emissions account for less than 5 percent of the world total.

In addition to the concentration of CDM projects in super-developing countries and the predominance of the non-CO_2 type projects, other emerging issues of the CDM have been criticized not only by developing countries but also by developed countries. In order to reply and address these criticisms in a practical manner, the CDM EB launched to ask a wide range of CDM stakeholders and the DNAs to input comments on efficiency in the operation of the CDM and opportunities for improvement.[7] As a result, the CDM EB received more than 40 comments mainly including issues concerning the demonstration of additionality and regional distribution matters. Concerning demonstration of additionality, the design of the CDM is likely to be inflexible for certain types of project activity, such as energy efficiency improvements, since the current design puts an over-emphasis on demonstration of the "additionality" of projects by investment analysis. In terms of the investment analysis, non-CO_2 type projects are increasingly advantaged because they have no or limited income other than emission credits. On the other hand, energy efficient improvement projects have to prove whether or not these issues would be a problem in the absence of emission credits by investment analysis.

The issue of regional distribution is the most serious and fundamental not only for the present CDM framework but also for its future design. As mentioned above, the projects are concentrated in super-developing countries such as China, India, and Brazil. However, as shown in Figure 3.3, regional distribution of the projects was more serious because it was clear that the projects were concentrated in Asian and South American countries with high economic activities represented by GDP and high investment opportunities from developed countries represented by Foreign Direct Investment (FDI). On the other hand, there are very limited projects in African and small island countries with comparably low economic activities and high dependency on official development assistance (ODA). Based on these facts, we can see that

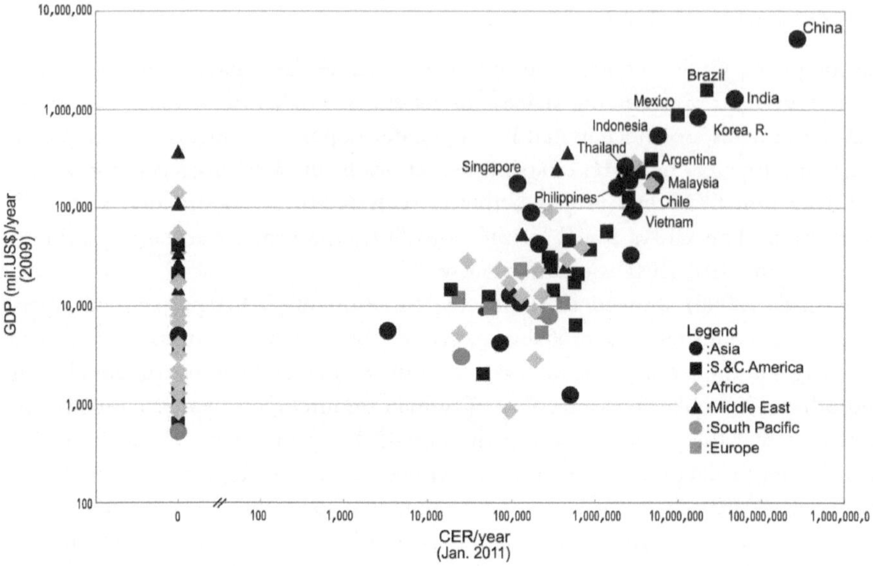

FIGURE 3.3 Relationship between estimated CERs and GDP (source: authors).

one of the purposes of the CDM – to assist in accomplishing the GHG reduction targets of Annex I Parties – has been achieved because of the issued 550 million CERs for developed countries; however, the other purpose – to contribute to the sustainable development of non-Annex I Parties – has not been achieved yet.

Reforming the CDM

The CDM EB has launched to reform the present CDM framework by adopting the management plan including five specific objectives to be achieved over the next two years at the 59th meeting.[8] In order to address the regional distribution matters, CDM EB has decided to include necessary works related to the geographic expansion of the CDM in the five objectives. The planned works are as follows:

a. establishment of an advisory forum on regional distribution;
b. assessment and, where possible, removal of barriers to the implementation of CDM projects and programmes in the CDM pipeline, in currently under-represented regions;
c. arrangements to ensure the operationalization of the loan scheme,[9] communicate its availability, and appropriately assist potential applicants to access funds;
d. top-down development of standards and standardized baselines that are particularly applicable in geographical regions, which are under-represented in the CDM;
e. strengthening DNA capacity and skills, and through the DNA Forum, to allow DNAs to prioritize, focus, and intensify their provision of support to assist stakeholders to capitalize on the benefits of the CDM;

f. effective coordination of the work programme of the Nairobi Framework partner agencies,[10] including their capacity-building activities to promote and improve regional distribution of CDM projects.

The authors consider that these works are essential to reform the present CDM framework, but additional works will be required in order to achieve the unrealized purpose of the CDM, which is to contribute to the sustainable development not only of super-developing countries but also of the countries under-represented in the CDM. In other words, considering criticism of the present framework of the CDM, a design for realizing another purpose concerning sustainable development will be inevitable to survive the CDM in the new regime after 2013.

What kind of CDM framework will make it possible to implement projects contributing to sustainable development, even in the least developed countries? First, the framework should at least have simplified procedures and modalities so that countries whose human resources are limited can develop and implement the projects. In order to simplify the procedures and modalities, in addition to the above-mentioned works proposed by the CDM EB, it is inevitable that drastic reform such as an abolishment of validation[11] to the projects under a certain size will become necessary. Second, but most important, is for the framework to have a system to secure financing for the projects. Until now, securing financing for not-so-profitable CDM projects has been difficult and many projects have been canceled even after registration by the CDM EB. The authors propose that a mechanism enabling secure financing and an evaluation of the contribution for the purpose of attaining sustainable development be integrated in the framework of procedures and modalities in the CDM.

Sustainable development criteria – bottom-up approach

Since "sustainable development" has a huge variety of models that depend on natural, social, and economic characteristics as well as the development level of each country, it is difficult to set uniform criteria to evaluate it. Therefore, a bottom-up approach developed by each country, considering their own characteristics should be adopted to evaluate the contribution of a project for sustainable development and compared to a top-down approach such as setting uniform criteria.

Scoring methodology to evaluate "co-benefit"

More precisely, a scoring method to evaluate the degree of contribution by the CDM project should be developed in each sector in the country. This contribution other than GHG emission reduction is recently considered as a "co-benefit." There are many co-benefit type GHG mitigation projects addressing stable electric power supplies, water and air pollution control, waste management issues, and poverty alleviation, etc. If the project has many co-benefits in each sector, the benefits should be taken into scoring consideration.

Co-benefits and financing system

Finally, the results of the scoring of the co-benefits should be in conjunction with the system to secure financing. The system may set a "premium" to GHG emissions reductions and/or a "priority" to be financed by developed countries or international development banks.

Partly because of the "inefficiency" of the current crediting system, the Japanese government is considering a "bilateral crediting mechanism" by which donor countries bilaterally certify a credit with a host country. A practical example would be a high-scoring co-benefit type GHG mitigation project prioritized in the nationally appropriate mitigation actions (NAMAs) plan prepared through the technical cooperation of JICA. Realized by yen loan, this could very likely stimulate CDM reform.

Notes

1 MOEJ (Ministry of the Environment) (2010) *CDM/JI Manual for Project Developers and Policy Makers 2010*, MOEJ, Japan.
2 The Methodology Panel (MP) in case of large-scale CDM methodology or small-scale CDM Working Group (SSC-WG) in case of small-scale CDM methodology considers their appropriateness of proposed methodologies, and recommends their approval to the CDM EB if their logicalities are fulfilled. Finally, the CDM EB decides whether they are approved or not.
3 Example from Thailand.
4 CDM Modalities and Procedures (CDM M&P), para. 44.
5 CDM M&P, para. 43.
6 CDM M&P, para. 63.
7 Call for inputs on efficiency in the operation of the CDM and opportunities for improvement (http://cdm.unfccc.int/public_inputs/2009/cdmimprov/index.html).
8 CDM MANAGEMENT PLAN 2011 (Version 01.1) (http://cdm.unfccc.int/EB/archives/meetings_10.html#59).
9 The Conference of the Parties serving as the meeting of the Parties to the Kyoto Protocol at sixth session (CMP6) adopted the guidelines and modalities for operationalization of a loan scheme to support the development of CDM projects in countries with fewer than ten such activities registered.
10 The Nairobi Framework was initiated by UNDP, UNEP, World Bank Group, African Development Bank, and the Secretariat of UNFCCC with the specific target of helping developing countries, especially those in sub-Sahara Africa, to improve their level of participation in the CDM.
11 In this case, project participants will submit the documents stipulated by the CDM EB. After receiving the documents, the UNFCCC secretariat will only check the contents, and if the contents of submitted documents are acceptable according to the manual provided by the CDM EB, the project will be automatically registered as a CDM project.

Reference

Yamada, Kazuhito (2008), "Institutional Design of Clean Development Mechanism (CDM)," doctoral thesis of Ibaraki University, pp. 184.

4

REDD+: A NEW STRATEGY FOR CONSERVING TROPICAL FORESTS

Scientific background of REDD+ and international discussions

Mitsuo Matsumoto

Introduction

The extensive destruction of rich, valuable tropical forests began in the 1970s, and calls for conservation soon followed. Despite warnings of the critical nature of the situation and attempts to control illegal logging, the destruction of tropical forests continues apace. Recently, a new program for protecting tropical forests is being discussed internationally, with the aim of mitigating climate change. The program is called "REDD+ (REDD-plus)" and was discussed also at COP15 (the Fifteenth Session of the Conference of Parties to the United Nations Framework Convention on Climate Change) held in Copenhagen in December 2009. This chapter summarizes the background and present situation of REDD+.

What is REDD+?

REDD+ is the abbreviation of a program for mitigating climate change, called "Reducing Emissions from Deforestation and Forest Degradation and the Role of Conservation, Sustainable Management of Forests and Enhancement of Forest Carbon Stocks in Developing Countries," which is being discussed in the Conferences of Parties to the United Nations Framework Convention on Climate Change (COP). The program is called either REDD+ or REDD-plus. Of REDD+ activities, "REDD" means those for reducing emissions from deforestation and forest degradation, while "+" denotes those for maintaining and increasing carbon stocks in conserved forests.

Today, international debate on REDD+ is based on the concept of positive incentives, which involves giving economic remuneration (incentives) for reductions in greenhouse gas emissions or increases in carbon stock achieved by reducing deforestation and forest degradation (Figure 4.1). The socioeconomic principle is

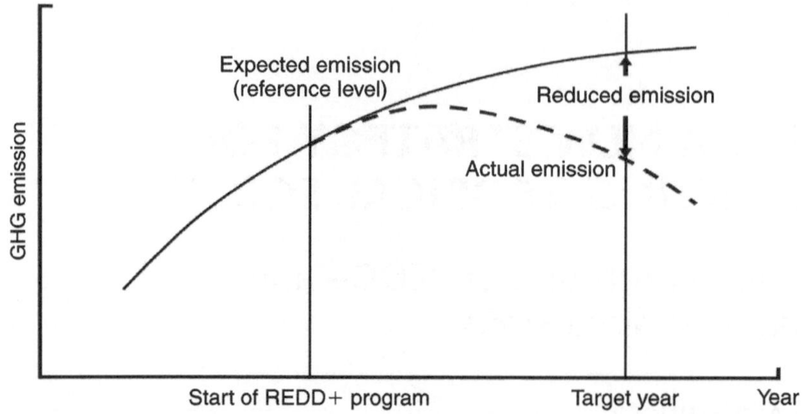

FIGURE 4.1 Concept of emission reductions in REDD+ (source: author).

that forests will be protected if it is more economically beneficial to maintain forests than to turn them into farmland, pasture, and cities, causing deforestation and forest degradation in the process.

Scientific background of REDD+

The first key to understanding the scientific background of REDD+ is reducing emissions from deforestation. According to the IPCC Fourth Assessment Report (2007a), approximately 80 percent of CO_2 emissions are from fossil fuels and cement, with the remaining 20 percent or so from changes in land use or deforestation (Figure 4.2). The Food and Agriculture Organization (FAO) showed that

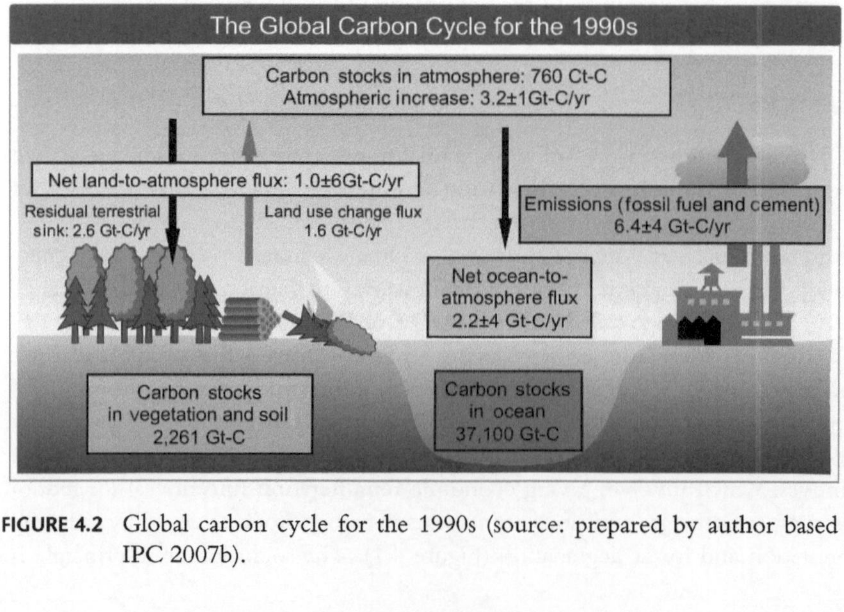

FIGURE 4.2 Global carbon cycle for the 1990s (source: prepared by author based on IPC 2007b).

deforestation is particularly serious in Brazil (Amazon), Indonesia, and Africa (FAO 2010). Total global deforestation was 8.3 million ha/year in 1990 to 2000, and 5.2 million ha/year in 2000 to 2010 (Table 4.1). The areas in Brazil and Indonesia combined accounted for 58 percent of global deforestation in 1990 to 2000, with 2.9 million ha/year and 1.9 million ha/year, respectively (Table 4.2). The area in Indonesia was decreased to 0.50 million ha/year in 2000 to 2010. In contrast, forest areas in China, India, and Vietnam were increasing from 2005 to 2010, reflecting a recent policy of promoting forestation. The present situation of changes in forest area varies greatly among developing countries.

The Stern Review (2007) stated that reducing deforestation is a highly cost-effective way of reducing greenhouse gases. The same view was also indicated in the IPCC Fourth Assessment Report (2007b), which stated that 65 percent of the potential for reducing costs below 100 US$ per ton of CO_2 is in the tropics, half of which can be achieved by reducing deforestation. Figure 4.3 shows that the cumulative carbon mitigation benefits by 2050 for a scenario of 2.7 US$/t$CO_2$+5 percent annual carbon price increment for one model are estimated to be 91,400 MtCO_2; 59 percent of it coming from avoided deforestation. It indicates also that during the period 2000–2050, avoided deforestation in South America and Asia

TABLE 4.1 Annual change in forest area by region and subregion, 1990–2010

Region/subregion	*1990–2000*		*2000–2010*	
	1,000 ha/yr	*%*	*1,000 ha/yr*	*%*
Eastern and Southern Africa	–1,841	–0.62	–1,839	–0.66
Northern Africa	–590	–0.72	–41	–0.05
Western and Central Africa	–1,637	–0.46	–1,535	–0.46
Total Africa	–4,067	–0.56	–3,414	–0.49
East Asia	1,762	0.81	2,781	1.16
South and Southeast Asia	–2,428	–0.77	–677	–0.23
Western and Central Asia	72	0.17	131	0.31
Total Asia	–595	–0.10	2,235	0.39
Russian Federation	32	n.s.	–18	n.s.
Europe excluding Russian Federation	845	0.46	694	0.36
Total Europe	877	0.09	676	0.07
Caribbean	53	0.87	50	0.75
Central America	–374	–1.56	–248	–1.19
North America	32	n.s.	188	0.03
North and Central America	–289	–0.04	–10	–0.00
Total Oceania	–41	–0.02	–700	–0.36
Total South America	–4,213	–0.45	–3,997	–0.45
World	–8,327	–0.20	–5,211	–0.13

Source: FAO (2010).

TABLE 4.2 Ten countries with largest annual net loss of forest area, 1990–2010

Country	*Annual change 1990–2000*		*Country*	*Annual change 2000–2010*	
	1,000 ha/yr	*%*		*1,000 ha/yr*	*%*
Brazil	–2,890	–0.51	Brazil	–2,642	–0.49
Indonesia	–1,914	–1.75	Australia	–562	–0.37
Sudan	–589	–0.80	Indonesia	–498	–0.51
Myanmar	–435	–1.17	Nigeria	–410	–3.67
Nigeria	–410	–2.68	United Republic of Tanzania	–403	–1.13
United Republic of Tanzania	–403	–1.02	Zimbabwe	–327	–1.88
Mexico	–354	–0.52	Democratic Republic of the Congo	–311	–0.20
Zimbabwe	–327	–1.58	Myanmar	–310	–0.93
Democratic Republic of the Congo	–311	–0.20	Bolivia (Plurinational State of)	–290	–0.49
Argentina	–293	–0.88	Venezuela (Bolivarian Republic of)	–288	–0.60
Total	–7,926	–0.71	Total	–6,040	–0.53

Source: FAO (2010).

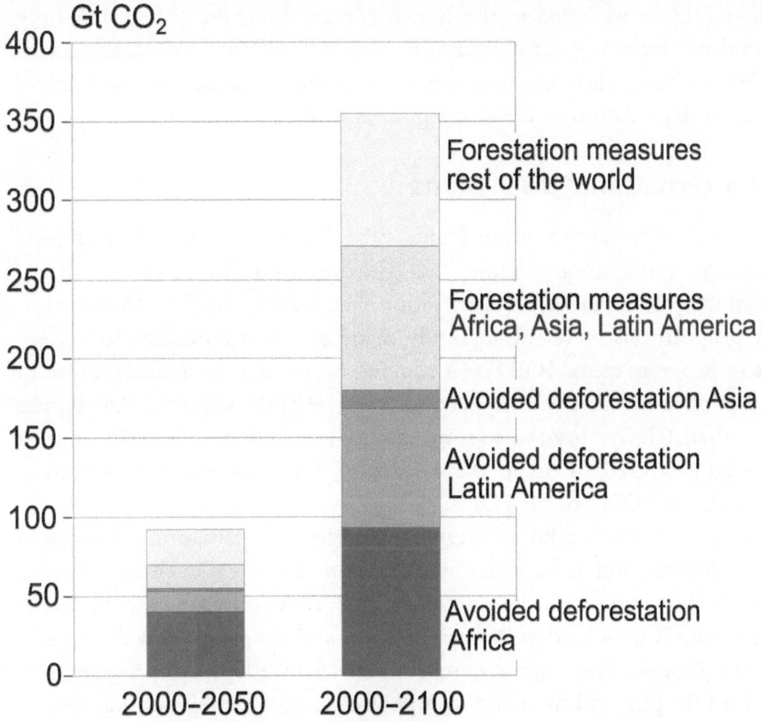

FIGURE 4.3 Cumulative mitigation potential (2000–2050 and 2000–2100) according to mitigation options under the 2.7 US$/t$CO_2$ +5%/yr annual carbon price increment (source: prepared by author based on IPCC 2007b).

dominate accounting for 49 percent and 21 percent, respectively, of the total mitigation potential, and that the mitigation potential of the continents Asia, Africa, and Latin America dominates the global total mitigation potential for the periods to 2050 and to 2100. Therefore, reducing deforestation and forest degradation is essential from a scientific viewpoint and is also likely to be less costly than other measures from an economic viewpoint. These facts are the basis of promoting REDD+.

The effects of REDD+ are not limited to mitigating climate change. For example, natural tropical forests are rich in biological diversity and provide numerous benefits, including forest products, water resources, and a richer environment. REDD+ is also expected to maintain these ecological services through forest conservation.

Although the need and economic benefits of reducing deforestation and forest degradation are widely recognized, no strategy was included in the Kyoto Protocol, which concerns international commitments related to climate change. This is because the Kyoto Protocol is intended to encourage advanced countries to reduce emissions, and there are no commitments or frameworks for developing countries on reducing emissions and REDD+. Only CDM by afforestation and reforestation, known as AR-CDM, is included in the Kyoto Mechanisms, but this covers only afforestation and not reducing deforestation or forest degradation. With this background, REDD+ is being discussed as a system for reducing emissions from deforestation and forest degradation in developing countries.

Progress of international discussions

The debate on REDD+ started when Papua New Guinea and Costa Rica proposed a measure for reducing greenhouse gas emissions by reducing deforestation, based on the motto of "Avoiding Deforestation," at COP11 held in Montreal in 2005. Similar propositions were subsequently made by other countries, and were integrated into a program called REDD (Reducing Emission from Deforestation in Developing Countries). Initially, the second D of REDD denoted developing countries, but others felt that it was also important to reduce forest degradation, and so it was decided at COP13 held in Bali to denote forest degradation as well as deforestation with the "DD" of "REDD."

The Bali Action Plan adopted at COP13 includes "conservation, sustainable management of forests, and enhancement of carbon stocks" as well as reducing emissions from deforestation and forest deterioration. This concept was included in the plans adopted at COP14 held in Poznan in 2008, and the program was renamed REDD+. In the Copenhagen Agreement at COP15 in 2009, the program was referred to as REDD-plus and the need to urgently create a framework was recognized (UNFCCC COP 2009).

The discussions on REDD+ initially focused on ways to avoid deforestation and moved to include forest deterioration, and the scope of protection was gradually extended to the present form of REDD+. Forest degradation was included because many developing countries, with the exception of Brazil, Indonesia, and tropical African countries, could not participate in the framework if the target was only deforestation. The "+" was added to include developing countries that had already implemented forest conservation or afforestation policies. Discussions progressed through multiple negotiations among countries with their own considerations and opinions, focusing on the relationship between the environment and economy.

Topics of REDD+

The topics for implementing REDD+ are described below from the two main perspectives of methodology, such as methods for monitoring and reference levels, and policy, including frameworks and funds for promoting REDD+.

Methodology

For the REDD+ program to give incentives depending on emissions reductions, it is important to appropriately monitor changes in carbon stocks in forests. The Subsidiary Body for Scientific and Technological Advice, thirty-first session (SBSTA31), which was held concurrently with COP15, advocated the need for building robust and transparent methods for monitoring forests. It was also decided to use the IPCC guidelines as estimation methodologies and to estimate changes in carbon stocks by combining remote sensing and ground surveys (UNFCCC, 2009).

The concept of monitoring methods is illustrated as follows using Figure 4.4. Analysis of remotely sensed images gives forest-type maps and the areas of each forest type. In contrast, ground surveys provide the carbon density of each forest type. Carbon stocks in a forest type are estimated by multiplying forest area (ha) with carbon density (C-t/ha). Here, carbon density indicates how much weight of carbon stocks by forest biomass exists in a unit area such as one hectare. Total carbon stocks are estimated by the total sum of carbon stocks in all forest types. Deforestation is monitored as changes of forest areas. Impacts of forest degradation and other plus activities are monitored as changes of carbon density. Therefore, monitoring changes of carbon density is necessary for evaluation of forest degradation and plus activities.

IPCC *Good Practice Guidance* (IPCC 2003) and the 2006 guidelines (IPCC 2006) indicate two methods for estimating carbon stock changes in forest biomass (Figure 4.5). One is the "default method" in the *Good Practice Guidance*, which is called the "gain-loss method" in the 2006 guidelines. With the default method, carbon stock changes in forest biomass are estimated by differences in carbon gain through

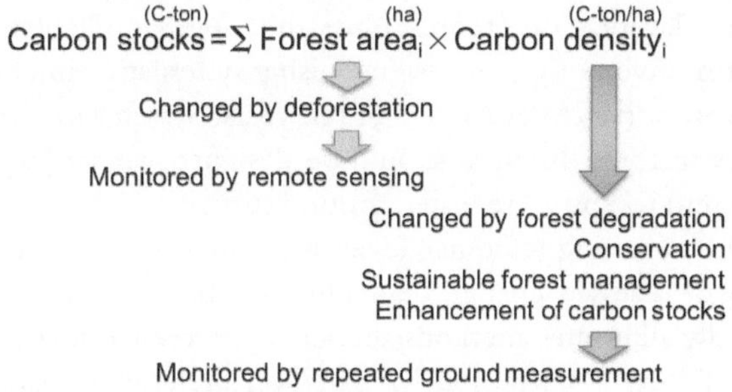

FIGURE 4.4 Basis of estimation methods of forest carbon stocks for REDD+ (source: author).

Default Method (gain–loss method)

Carbon stock changes
= Gain by growth – Loss by disturbance
↓
Not easy to estimate

Stock Carbon Change Method (stock–deference method)

Carbon stock changes
= (Carbon stocks$_{t2}$ – Carbon stocks$_{t1}$)/(t2 – t1)

gives feasible estimation
requires Repeated Ground Measurement

FIGURE 4.5 Estimation methods of carbon stock changes in forest biomass (source: author).

growth and carbon loss through human induced and natural disturbances. Although the method is basic and applicable for general purposes, estimates of losses through disturbances are not easy because statistics and data are insufficient for illegal harvesting and natural disturbances such as fires and winds. Another method of estimating carbon stock changes in forest biomass is the "stock change method" in the *Good Practice Guidance*, which is called the "stock-difference method" in the 2006 guidelines. With the stock change method, carbon stock changes are estimated based on differences in carbon stocks in repeated surveys. While the stock change method is more feasible than the default method because an estimation of carbon loss is not required, repeated ground surveys are necessary.

Based on the concept, Kiyono *et al.* (2010) estimated carbon stocks in forest biomass in Cambodia at a national level. They made a forest map by analysis of satellite images to estimate forest areas of two main species including evergreen forests and deciduous forests. On the other hand, they estimated averaged carbon densities of the forest types from survey of 155 permanent sample plots using allometry equations. Then they estimated total carbon stocks in Cambodia to be 863.0 ± 403.3 Mt-C (Table 4.3). The next assessment of carbon stocks will give carbon stock changes using the stock change method.

The uncertainty of the estimation in Cambodia was calculated at 47 percent. To reduce the uncertainty, more classification of forest types and more sample plots are necessary. Most national forest inventories are designed using systematic samplings, and permanent sample plots are surveyed every five or ten years. Systems for monitoring carbon stock changes in forest biomass should be designed, respecting the national forest inventory system to ensure repeated ground surveys.

Besides monitoring methods, setting reference levels is an important methodology, because reductions to be assessed change depending on the reference level used. Figure 4.6 schematically illustrates methods for setting reference levels and reductions to be assessed. In Figure 4.6, Method 1 is based on a reference year, and uses the same concept as that of the reference year used in the Kyoto Protocol. The advantages and disadvantages are strongly expressed when annual fluctuations in emissions are large. In Method 2, which is based on a reference period, annual fluctuations are leveled off. Method 3 uses a simple model, which shows changes in emission with a linear regression equation, and is relatively simple to use but may

TABLE 4.3 Results of a preliminary estimation of forest carbon stocks in Cambodia

Forest type	*Forest area (km^2)*	*Averaged carbon density (t-C/ha)*	*Total carbon stocks (Mt-C)*
Evergreen	36,689	140.0 ± 89.5	513.6 ± 328.3
Deciduous	46,921	74.5 ± 49.9	349.4 ± 234.3
Total	83,610		863.0 ± 403.3 Uncertainty: 47%

Source: prepared by the author based on Kiyono *et al.* (2010).

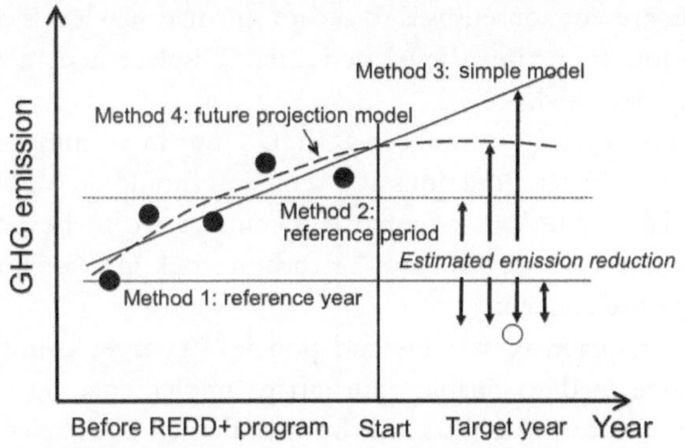

FIGURE 4.6 Comparison of methods of setting reference levels and emission reductions to be assessed (source: author).

give extreme values such as that shown in the figure. There is also the question of whether it is possible to obtain estimates of past emissions sufficient to prepare a regression equation. Method 4 uses a future projection model. It involves developing a model for estimating emissions using explanatory variables such as GDP and farmland area, projecting future emissions using the model, and setting the reference level. It requires a large volume of historical data to be collected, and applicable countries are likely to be limited.

As described above, methods for setting reference levels affect the amount of incentives given. Various ideas have been proposed and discussed, but nothing specific has been debated in international discussions. Future practical implementations and comparative studies are required for more thorough discussions.

Policy

The main policy related topics of REDD+ include frameworks, funding sources, boundaries, MRV (measurement, reporting, and validation), and domestic policies. What framework is to be constructed and how it is to be funded are the essence of REDD+. From the outset, discussions have examined whether the program should be funded by international funds or by market credits. Recently, one idea gaining support involves fostering the capacities of developing countries by using international funds at the initial stage, paying limited expenses out of international funds at the second stage, and finally moving to the full-scale market credit method. This method is called a phased approach and is used in numerous discussions.

Since the start of discussions, regional boundaries at REDD+ have been national borders, because setting boundaries to create many small areas could lead to the displacement of emissions from one area to another. Here, displacement is equivalent to "leakage" in the Clean Development Mechanism (CMD) of the Kyoto Protocol and means increasing emissions outside a boundary to reduce emissions inside the boundary. In today's negotiations, some countries claim that sub-national levels should be permitted because it is difficult to perform activities and monitor forests at national levels. The present consensus is to accept sub-national levels only tentatively until full transition to national-level activities. Displacement across national borders is also being discussed.

MRV is a widely used abbreviation not only in REDD+ but in all mitigation activities. The concept is that emission reductions and removals should be Measurable, Reportable, and Verifiable. MRV is essential if incentives are to be given according to the volume of emissions reduced and/or carbon stock increased, and so the term is used to convey the concept.

To implement REDD+ in practice, systems and policies of target countries should be constructed or reformed to enable appropriate implementation. For example, a new system for monitoring forests will be meaningless if policies for reducing deforestation or forest degradation are not taken. In the program, causes of deforestation and forest degradation in the country should first be determined, and then measures should be decided and taken. According to the SBSTA31 agreement on methodology (UNFCCC 2009), target countries are to identify drivers of deforestation and forest degradation and determine activities that will reduce emissions, increase removals, and stabilize the carbon stock in forests. National governance is indispensable in REDD+, and designing systems that encourage appropriate governance is important when designing the REDD+ framework.

In international negotiations on REDD+, topics on biological diversity and the rights of indigenous and local people are being discussed as safeguards. There are strong expectations for REDD+ from an ecological point of view, such as for conserving biodiversity. Closer liaison with the Convention on Biological Diversity could bring synergies. Although there has been no discussion on how to incorporate biodiversity aspects into REDD+, proposals are requested based on past projects. A great deal of time and effort has been spent on the rights of indigenous and local people. The written decision of COP15 also mentions the need to involve indigenous and local communities in measuring and reporting activities related to REDD+.

Present state of REDD+

Although international discussions on REDD+ have made progress, the United Nations Framework Convention on Climate Change (UNFCCC) has not decided how to build an international system. Nevertheless, a number of voluntary activities on REDD+ are already underway. At Oslo Forest Climate Conference 2010, "REDD+ Partnership" was formed by countries keen to promote REDD+ activities and support funding. This aims to promote REDD+, as the official agreement process has not been decided. There are also ongoing multinational and bi-national programs related to REDD+, such as the Forest Carbon Partnership Facility of the World Bank, UN-REDD by the Food and Agriculture Organization (FAO), United Nations Development Programme (UNDP), and United Nations Environment Programme (UNEP), and a partnership between Norway and Indonesia.

In the private sector, the Verified Carbon Standard (VCS), which is the first credit system of REDD+ activities, is already widely used in voluntary REDD+

projects. These voluntary frameworks are likely to affect the future frameworks and accounting rules in UNFCCC and should be closely monitored. Japan should also consider developing its own voluntary methodology and accounting rules.

Conclusions

This chapter summarized REDD+, its scientific background, and international discussions. REDD+ is attracting immense interest, although the end point remains unclear. Each country appears to have its own agenda and to be focused only on economic benefits, with little interest in forest issues. However, this is precisely why academia should actively participate in international discussions on REDD+ and remind the international community of the need for scientific approaches and for the conservation of forests and the global environment.

With this background, Japan began chairing the REDD+ Partnership and have initiated programs related to REDD+, such as Program Grant Aid for the Environment and Climate Change for "forest conservation," operated by the Ministry of Foreign Affairs. As a base in Japan, the Forestry and Forest Products Research Institute established the REDD Research and Development Center in July 2007 (www.ffpri.affrc.go.jp/news/2010/20100804redd/documents/redd.pdf). The center will promote forest conservation activities through the private and other sectors by conducting research related to REDD+, developing policies and monitoring methods, training and disseminating forest technologies, holding seminars, and distributing information.

References

FAO (Food and Agriculture Organization) (2010) *Global Forest Resources Assessment 2010*. Rome: FAO.

IPCC (Intergovernmental Panel on Climate Change) (2003) *Good Practice Guidance for Land Use, Land-Use Change and Forestry*, J. Penman, M. Gytarsky, T. Hiraishi, T. Krug, D. Kruger, R. Pipatti, L. Buendia, K. Miwa, T. Ngara, K. Tanabe, and F. Wagner (Eds.). Hayama, Japan: Institute for Global Environmental Strategies (IGES).

IPCC (Intergovernmental Panel on Climate Change) (2006) *Volume 4: Agriculture, Forestry and Other Land Use*, S. Eggleston, L. Buendia, K. Miwa, T. Ngara, and K. Tanabe (Eds.). Hayama, Japan: IGES.

IPCC (Intergovernmental Panel on Climate Change) (2007a) *Climate Change 2007: The Physical Science Basis. Contribution of Working Group I to the Fourth Assessment Report of the Intergovernmental Panel on Climate Change*, S. Solomon, D. Qin, M. Manning, Z. Chen, M. Marquis, K.B. Averyt, M. Tignor and, H.L. Miller (Eds.). Cambridge and New York: Cambridge University Press, 996pp.

IPCC (2007b) *Climate Change 2007: Mitigation of Climate Change. Contribution of Working Group III to the Fourth Assessment Report of the Intergovernmental Panel on Climaate Change*, B. Metz, O.R. Davidson, P.R. Bosch, R. Dave, and L.A. Meyer (Eds.). Cambridge and New York: Cambridge University Press, 851pp.

Kiyono, Y., Furuya, N., Sum, T., Umemiya, C., Itoh, E., Araki, M., and Matsumoto, M. (2010) "Carbon Stock Estimation by Forest Measurement Contributing to Sustainable

Forest Management in Cambodia," *JARQ (Japan Agricultural Research Quarterly)*, 44 (1): 81–92.

Stern, N. (2007) *Stern Review on the Economics of Climate Change.* New York: Cambridge University Press, 692pp.

UNFCCC (United Nations Framework Convention on Climate Changes) (2009) "Reducing Emissions from Deforestation in Developing Countries: Approaches to Stimulate Action," Document FCCC/SBSTA/2009/L.19/Add.1, Subsidiary Body for Scientific and Technological Advice, Copenhagen, Denmark, December 8–12, 2009, 3pp.

UNFCCC COP (2009) Copenhagen Accord, UNFCCC, 2pp.

5

INTERNATIONAL INTERCITY COLLABORATION AND ITS POTENTIAL APPLICATION IN LOW CARBON DEVELOPMENT

Hidenori Nakamura and Hideyuki Mori

Introduction of international intercity collaboration for sustainable development

This chapter introduces the international intercity networks, or platforms, for sustainable cities in Asia, and discusses potential application of the network in climate change mitigation through strengthening local-level management of low carbon development.[1] An international intercity platform is desirable in order to facilitate local low carbon development initiatives, especially in developing countries. Such collaborative activities among cities have the potential to produce effects that may not happen without collaboration and networking: dissemination and transfer of knowledge, policies, and practices; development of joint projects in business and research; and increasing the international presence of the city (Nakamura *et al.* 2010a; Fujikura *et al.* 2010). The High Level Seminar on Environmentally Sustainable Cities (ESC), held in Jakarta, Indonesia in March 2010 at the East Asia Summit Environment Ministers Meeting, also confirmed that forming ESC was a priority topic in Asia. Strengthening support for intercity networks was also recommended.[2]

This chapter introduces the objectives and activities of international intercity collaboration for sustainable development. Lessons learnt from past and current collaborations are reviewed and examined in traditional environmental management fields, such as pollution control and waste management, as well as climate change mitigation. Strategies for formulating and managing international intercity collaboration are also discussed based on past experience and relevant studies on knowledge management for sustainable development. As an extension of past and current intercity collaboration on traditional environmental management, we then propose an Asian intercity platform for low carbon development, together with four potential activities and the expected role of the secretariat.

In starting a discussion of international intercity collaboration, it is better to first examine the definition of international intercity collaboration for sustainable development, the broad characteristics of which are outlined below.

- International intercity networks, or relationships of city stakeholders, are generated and maintained mostly via meetings of stakeholders from more than one country who are engaged in a city's sustainability policies and technologies; sometimes this also uses Internet communications.
- The networks could be institutionalised as programmes.
- The basis, or common foundation, of the networks is to develop new relationships, actions, policies, products, knowledge, and so on, which could contribute to the promotion of sustainable cities.
- Some new benefits are expected to come out of the collaboration, which would be beneficial to the participants. Positive externalities from the network are anticipated, as exemplified by some intercity networks and studies of knowledge networks: CITYNET's "Together, we can do more"; and the point noted by Creech and Willard (2001) that the total is bigger and better than the sum of the individual parts.

Examples of international intercity collaboration for sustainable development include CITYNET, the International Council for Local Environmental Initiatives (ICLEI), and the Kitakyushu Initiative for a Clean Environment. According to a survey by the Clean Air Initiative for Asian Cities (CAI-Asia) and the Cities Development Initiative for Asia (CDIA) that surveyed 14 organisations with city networks that included Asian cities or worked closely with city networks, 11 organisations focused on urban development such as urban planning and governance, and ten organisations also focused on urban environment management such as solid waste management and climate change adaptation and mitigation (CAI-Asia and CDIA 2009). This survey found that all organisations provided technical assistance to participating cities, and around half of them provided funding to cities through grants. Network activities included networking and advocacy, research, assistance for formulation of plans and strategies, project implementation, and capacity development. Most of the networks were funded by donors, and in some cases by membership dues (CAI-Asia and CDIA 2009).

The objectives of international intercity collaboration for sustainable development include:

- developing joint international projects among stakeholders, such as starting new businesses and joint research;
- developing new policies, products, and knowledge, including knowledge creation through joint research and development;
- learning about knowledge, policies, and practices, through diffusion, dissemination, and transfer of knowledge; and
- promoting the city, or increasing the presence of the city.

In most cases the participants in such intercity collaboration are local governments, corporations, research institutions, or (potentially) civil society organisations. International intercity collaboration activities for sustainable development include: mutual reference and learning as well as capacity development; agreements and exchanges to promote environmental business; implementing joint projects; and policy advocacy. Last, regarding the institutional structure and organisation of international intercity collaborations for sustainable development, a range of examples exist from networks with a secretariat, which are very formal, to ad hoc relationships without a formal secretariat to serve as a permanent coordinator among members. Some intercity networks operate in the form of city-to-city cooperation while others are multilateral.

Lessons learnt from past experience

Existing Asian intercity network programmes for the environment: CITYNET, ICLEI, Kitakyushu Initiative, CAI-Asia, and others

In this subsection, several international intercity network programmes for the environment are introduced and their institutional settings are discussed. The examples of programmes provided here comprise programmes with a secretariat, and examples of the mutual reference achievements attained in activities under these network programmes are shown. The enabling factors for local governments to adopt and disseminate practices through intercity networks are also presented, followed by an analysis of motivations and incentives for cities to participate in and utilise network programmes. Expected and actual benefits of participation are also discussed.

Introduction to existing network programmes[3]

Four international environmental cooperation networks that focus on environmental management, and include local governments in cities in East Asia as the main actors in the network, are presented here, in order to introduce examples of international intercity network programmes:[4]

- the Kitakyushu Initiative for a Clean Environment (hereafter referred to as the Kitakyushu Initiative);
- the Clean Air Initiative for Asian Cities (hereafter referred to as CAI-Asia);
- ICLEI;[5] and
- CITYNET.

The Kitakyushu Initiative, CAI-Asia, ICLEI, and CITYNET all focus on a common goal: environmental improvement and sustainable development. These networks aim to contribute to this goal by carrying out international activities to share experience and improve capacity. The Kitakyushu Initiative, CAI-Asia, ICLEI, and CITYNET started their activities in 2000, 2001, 1990, and 1987, respectively.

The Kitakyushu Initiative, ICLEI, and CITYNET have local governments as their main participating members; other governmental agencies, international development organisations, local governmental associations, research institutes, and other organisations are also involved in their activities as cooperating organisations or associate members. CAI-Asia includes the activities of multi-stakeholders, including organisations other than local governments; in addition, local governments from developed countries do not participate in CAI-Asia.

The local governments that participate in these networks do not include all the pertinent municipalities in all the countries involved: rather, most of them are local governments that are carrying out advanced approaches to the environment. Local governmental participation is voluntary in ICLEI, CAI-Asia, and CITYNET; participation by local governments in the Kitakyushu Initiative was first determined on the recommendation of participating national governments at the Ministerial Conference for Environment and Development in Asia and the Pacific (MCED), but, at present, local governments that are interested in participating are also being included.

In all cases, activities gain the support of international development agencies, and pilot projects/demonstration projects are implemented with the support of donors. However, the local governments from developed countries participating in these networks are viewed as cooperating actors, and there are cases of local governmental cooperation in which cooperative activities have been carried out with cities in developing countries (CITYNET, Kitakyushu Initiative). ICLEI and CITYNET also collect fees from participating local governments and carry out activities with external support.

In addition, information about specific activities and support tools for capacity building by each of the four international networks presented in this study has been divided into the following three categories: face-to-face contacts (meetings to share knowledge and experience, seminars/workshops, training, study tours, dispatching experts/consultation); documentation (creating successful practice databases, developing and selling case study compendiums, carrying out research and studies, and developing and selling reports and manuals, etc.); and online materials (publishing documents and workshop materials on websites, and providing activity support tools).

Table 5.1 presents the activities of the four networks arranged according to activities corresponding to changes and effects initiated by the networks. All the networks provide financial and technical cooperation, as well as carrying out capacity-building activities; however, some networks (and not others) promote support for creating domestic systems and local governmental cooperation. CAI-Asia and CITYNET and the Kitakyushu Initiative support the creation of domestic systems, and CITYNET promotes cooperation between local governments. Details on the individual activities of each network are provided in Tables 5.2 through 5.5.[6]

The Kitakyushu Initiative has achieved results in demonstration projects through financial and technical cooperation, by providing grants of US$3,000 to 10,000 from the United Nations to implement project formulation studies on urban

TABLE 5.1 Breakdown of activities of four international environmental networks

Activity area	*Kitakyushu Initiative*	*CAI-Asia*	*ICLEI*	*CITYNET*
Financial and technical cooperation	✓	✓	✓	✓
Capacity-building activities	✓	✓	✓	✓
Domestic institutional development	–	✓	–	✓
Promotion of bilateral intercity cooperation	✓	–	–	✓

Source: Nakamura *et al.* (2010b).

TABLE 5.2 Kitakyushu Initiative activities

Activity area	*Details*
Financial and technical cooperation	Demonstration projects – financial assistance of US$3,000 to 10,000 from the United Nations Economic and Social Commission for Asia and the Pacific, which is the main organiser of the Kitakyushu Initiative, to cities participating in the network for implementing new measures regarding air, water, waste, energy, and other areas; this financial assistance is used for project development studies, guidance from experts, procuring necessary equipment, and conducting seminars. Eleven projects were implemented together with financial support from implementing cities.
Capacity-building activities	Network meetings to share knowledge and experience Thematic seminars Study tours Studies and research by the secretariat on specific topics Created a database on successful practices Publishes documents and workshop materials on website Publishes newsletters
Promotion of bilateral intercity cooperation	Cooperation mediation by secretariat

Source: Nakamura *et al.* (2010b).

environmental issues, provide expert guidance, procure required materials and equipment, and conduct seminars. In addition to organising thematic seminars and study tours, the network also offers a database on the Internet as a capacity-building activity. The secretariat also mediates cooperation between cities (see Table 5.2).

Pilot projects are carried out by CAI-Asia in the form of financial and technical cooperation, and mainly focus on studies, with support from international development organisations such as the Asian Development Bank (ADB). In addition to the development of training manuals and country-specific integrated reports on urban

TABLE 5.3 CAI-Asia activities

Activity area	*Details*
Financial and technical cooperation	Pilot projects – studies on air pollution and health effects through financial assistance from the ADB and health damage research labs, US$800,000 in total; studies on investment planning formation for sustainable transportation through financial assistance from the Swedish International Development Agency, SIDA, and the ADB; technical cooperation to reduce air pollution through the use of diesel vehicles through financial assistance from the World Bank and the United States Agency for International Development, USAID, etc.
Capacity-building activities	Workshops/policy forums Training courses Develops training manuals, air quality improvement projects/ study compendiums, and integrated reports on urban air quality management by country, etc. Publishes documents and workshop materials on website Online discussion forum (listserv) and mailing list
National support system	Supports developing domestic networks (local networks) with exclusive secretariats

Source: Nakamura *et al.* (2010b).

TABLE 5.4 ICLEI activities

Activity area	*Details*
Financial and technical cooperation	Capacity improvement project using urban environmental management tools for local governments (supporting organisation: European Union) Project on stakeholder evaluation and adoption of agendas for integrated water management (ADB) Project on introducing measures for climate protection in the fields of renewable energy, transportation, energy efficiency, and waste (USAID; Canadian International Development Agency (CIDA); and others)
Capacity-building activities	Workshops Created a database on successful practices Provides activity support tools on website Publishes newsletters

Source: Nakamura *et al.* (2010b).

TABLE 5.5 CITYNET activities

Activity area	*Details*
Financial and technical cooperation	Tsunami Reconstruction Support Project through support from Yokohama City Government, which is CITYNET's sponsor city Environmental education and technology cooperation project through support from JICA and Yokohama
Capacity-building activities	Seminars, workshops Training courses (related to continuous urban environmental services in the International Training Centre for Local Authorities in Kuala Lumpur, Malaysia) Dispatching experts/consultations Develops and publishes reports and manuals Publishes documents and workshop materials on website Publishes newsletters
National support system	Support for developing national networks and programmes (national chapters)
Promotion of bilateral intercity cooperation	Cooperation mediation by the secretariat Matching through website

Source: Nakamura *et al.* (2010b).

air quality management, CAI-Asia publicises reference documents on its website and manages an online discussion forum (listserv), as part of its capacity-building activities. In addition, support is provided for developing domestic networks (local networks) with exclusive secretariats (see Table 5.3).

ICLEI Southeast Asia provides financial and technical cooperation to support projects for introducing measures in the fields of renewable energy, transportation, energy efficiency, and solid waste for climate protection, with the support of the United States Agency for International Development, among others. In addition, ICLEI also carries out capacity-building activities, such as developing databases for workshops and successful practices, and provides support tools through the Internet (see Table 5.4).

Last, CITYNET carries out technical cooperation projects for environmental education through support from the Japan International Cooperation Agency (JICA) and the Yokohama City Government. In addition, CITYNET also conducts capacity-building activities through continuous training courses, dispatching experts, and providing materials on their website, and provides support for developing domestic network programmes (national chapters). The secretariat also mediates cooperation between cities on the web (see Table 5.5).

Mutual reference and learning

The evidence demonstrates that the three networks CITYNET, ICLEI Southeast Asia, and the Kitakyushu Initiative have actually led to additional bilateral intercity relationships being formed.[7] In the case of CITYNET, bilateral cooperation between two cities has come about through the support of the secretariat for the following pairs of cities: Yokohama (Japan) and Banda Aceh (Indonesia), Hue (Vietnam), Hanoi (Vietnam), Incheon (Korea), and Phnom Penh (Cambodia); Penang (Malaysia) and Dhaka (Bangladesh); Bangkok (Thailand) and Hue; and Seoul (Korea) and Makati (Philippines). The forms of the cooperation included study tours, feasibility studies, on-the-job training, exchanges of know-how, needs assessments, support for project formation, and training sessions.

The state of member cities' participation in network activities will be examined first. Figure 5.1 shows the frequency of participation in network events by member cities of CITYNET, ICLEI Southeast Asia, and the Kitakyushu Initiative, respectively, from highest to lowest. The numbers of cities that have participated in network events at least once for CITYNET, ICLEI Southeast Asia, and the Kitakyushu Initiative, are 63, 48, and 47, respectively. The numbers of events organised by CITYNET, ICLEI Southeast Asia, and the Kitakyushu Initiative are 67, 53, and 16, respectively. Therefore, the percentages of member cities that participated in at least one activity for CITYNET, ICLEI Southeast Asia, and the Kitakyushu Initiative were 90 per cent, 91 per cent, and 76 per cent, respectively, and thus the overall rate of participation seems high, although a few cities have never participated in any events. The periods up to 2008 during which programmes were offered for CITYNET, ICLEI Southeast Asia, and the Kitakyushu Initiative are 22, 13, and nine years, respectively, and thus the numbers of cities that participated in at least one event every two years, on average, for CITYNET, ICLEI Southeast Asia, and the Kitakyushu Initiative are 12, 14, and 18, respectively.[8] These cities can be seen to constitute a group that demonstrated some commitment to network

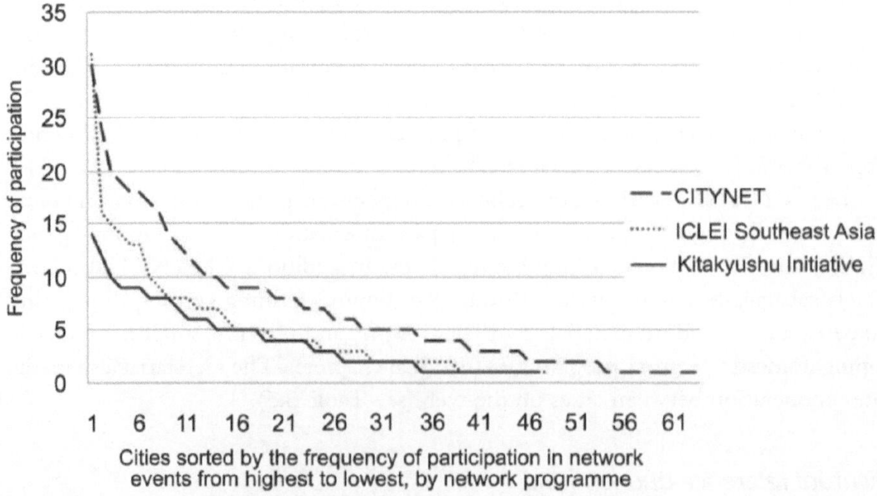

FIGURE 5.1 Distribution of the frequency of cities' participation in network events by network programme (source: Nakamura *et al.* 2010).

activities and contributed to maintaining the networks. The above analysis is summarised in Table 5.6.

As shown in Figure 5.1, the distribution of the frequency of cities' participation is similar in the programmes of all three networks, and it is asymmetrical. Some cities had a greater frequency of participation while most participating cities participated in network events more selectively. The five most actively participating cities for each network are shown in Table 5.7. A higher frequency of participation implies that a city had the intention to proactively learn from or contribute to knowledge dissemination at network events. This should be particularly true of cities from developing countries that frequently attend events. Cities that had a higher frequency of participation could be active learners. However, in interpreting these results, it is important to consider that some events may have been more or less significant than others. Just because a city participated in a limited number of events does not necessarily mean that mutual learning was not significant. It may be that some cities that participated in only a small number of network events might

TABLE 5.6 Frequency of cities' participation in network events

	CITYNET	*ICLEI Southeast Asia*	*Kitakyushu Initiative*
Number of participating cities	70	53	62
Years of network events	22	13	9
Total number of events over the network lifetime	67	53	16
Number of cities participating in network events at least once over the life of the network	63	48	47
Percentages of cities participating in network events at least once over the life of the network	90	91	76

Source: Nakamura *et al.* (2010b)

TABLE 5.7 Top five cities in terms of frequency of participation in network events

	CITYNET	*ICLEI Southeast Asia*	*Kitakyushu Initiative*
1	Yokohama (Japan)	Baguio (Philippines)	Kitakyushu (Japan)
2	Colombo (Sri Lanka)	Bohol Province (Philippines)	Surabaya (Indonesia)
3	Makati (Philippines)	Naga (Philippines)	Nonthaburi (Thailand)
4	Dhaka (Bangladesh)	Puerto Princesa (Philippines)	Dhaka (Bangladesh), Weihai (China), Bangkok (Thailand)
5	Kuala Lumpur (Malaysia), Kathmandu (Nepal)	Cebu (Philippines), Mandaue (Philippines)	–

Source: Nakamura *et al.* (2010b).

nevertheless have considered these events to be very significant for their learning. Cities that participated in many events may have found only a few to be significant, but the value of participating in these significant events was enough to justify continued participation in the network.

This uneven distribution of participation by member cities could be explained in two ways. First, it might reflect the fact that some cities from developing countries may be receiving relatively more financial assistance for participation. In some cases a network programme may allocate some financial resources to encourage active participation by certain member cities. Limited financial resources to invite cities would be allocated to cities that show an interest in active participation, and that have the potential to effectively utilise the network activities. Second, it may indicate that some cities that actively participated in many network events are "hub cities" in the networks. Hub cities could be from both developed and developing countries. These cities include Yokohama and Kitakyushu (Japan), Makati and Baguio (Philippines), Surabaya (Indonesia). Hub cities are either eager to learn from other cities, or to disseminate their own practices to other cities; some cities, especially from developing countries, are interested in both. The orientation of each member city can be directly observed from an analysis of the cities' participation in network activities, as explained in the following section.

It can be inferred that cities that actively participated in the network events believed that the network programmes were worthwhile, whereas inactive cities saw few benefits. Participation in these networks is costly for cities, not only in monetary terms, but also in terms of human resources. In order to participate in training, workshops, network meetings, and other activities that promote mutual learning, city officials need to take time away from their daily tasks. This is true for host cities as well. This cost is high for cities in developing countries in particular, which typically suffer from severe shortages of qualified staff. Therefore, if cities are willing to pay these costs, especially over a long period of time, it suggests that the cities believe that the benefits are worth the costs, and that mutual reference and learning is in fact taking place.

To be sure, mutual reference and learning may not be the only motivation for cities to actively participate in international intercity networks. Cities may also participate not just for material benefits such as knowledge or financing, but also for non-materialistic reasons of reputation and pride. Proactive cities with a large frequency of participation may build up a reputation among participating cities or even beyond the community of member cities. Such an enhanced reputation may lead to a sense of satisfaction and pride in local government officials, or to the city's leaders demonstrating their own capacity, which could contribute to their re-election and/or reappointment.

The records of cities' participation in network events also imply frequent mutual reference among several cities that often participated in the same events, in particular Colombo (Sri Lanka), Dhaka, Kuala Lumpur (Malaysia), Makati, and Bangkok. Mutual reference among three of these cities in particular (Colombo, Dhaka, and Makati) is considered to be truly mutual, meaning that these cities were learning and providing knowledge in both directions. The directions of intercity mutual reference are shown in Table 5.8 based on the records of bilateral intercity cooperation and records of joint participation by the six cities that most frequently participated in network events. This table shows cities that learnt certain environmental policies or measures, or received specific environmental cooperation support, according to the respective reference city or cities.

In the case of ICLEI Southeast Asia's programme Cities for Climate Protection (CCP), Baguio studied bio-fuel utilisation for climate change mitigation with Chiang Mai (Thailand), and Naga (Philippines) learnt about organic fertiliser production from waste in a partnership with Tungsong (Thailand). Records of cities' participation in network events imply frequent mutual learning among Baguio, Cebu (Philippines), Bohol Province (Philippines), Naga, and Puerto Princesa (Philippines). Among these cities, Baguio, Naga, and Puerto Princesa were also reference cities for the other cities. These intercity relations are shown in Table 5.9.

In the case of the Kitakyushu Initiative, relationships have developed through technical cooperation by dispatching experts and conducting training courses between Kitakyushu and Dalian (China), Kitakyushu and Surabaya, Kitakyushu and Kathmandu

TABLE 5.8 State of intercity mutual reference (CITYNET)

Cities referring to other cities	*Cities referred to by other cities*
Kathmandu (Nepal)	Colombo (Sri Lanka), Kuala Lumpur (Malaysia)
Dhaka (Bangladesh)	Colombo (Sri Lanka), Penang (Malaysia), Phnom Penh (Cambodia), Makati (Philippines)
Colombo (Sri Lanka)	Bangkok (Thailand), Kuala Lumpur (Malaysia), Makati (Philippines)
Banda Aceh (Indonesia)	Yokohama (Japan)
Hanoi (Vietnam)	Yokohama (Japan)
Hue (Vietnam)	Bangkok (Thailand), Yokohama (Japan)
Phnom Penh (Cambodia)	Makati (Philippines)
Makati (Philippines)	Seoul (Korea)
Incheon (Korea)	Yokohama (Japan)

Source: Nakamura *et al.* (2010b).

TABLE 5.9 State of intercity mutual reference (ICLEI Southeast Asia)

Cities referring to other cities	*Cities referred to by other cities*
Baguio (Philippines)	Chang Mai (Thailand)
Naga (Philippines)	Tungsong (Thailand), Baguio (Philippines), Puerto Princesa (Philippines)
Cebu (Philippines)	Baguio (Philippines), Naga (Philippines), Puerto Princesa (Philippines)
Puerto Princesa (Philippines)	Baguio (Philippines)
Bohol Province (Philippines)	Cebu (Philippines), Puerto Princesa (Philippines)

Source: Nakamura *et al.* (2010b).

(Nepal), Nonthaburi (Thailand) and Dhaka, Ulsan (Korea) and Kathmandu, and through network exchanges and mediation by the United Nations Economic and Social Commission for Asia and the Pacific (UNESCAP), the sponsor of the network, and the Institute for Global Environmental Strategies (IGES), the network secretariat. In addition, information collection between cities is also being carried out, as seen in the case of Bago (Philippines), which has been collecting information regarding waste treatment, particularly composting and recycling, from Surabaya and Bangkok. The record of cities' participation in network events also implies active bilateral referencing among Kitakyushu, Surabaya, Nonthaburi, Dhaka, Weihai (China), and Bangkok. Among these cities, Kitakyushu, Surabaya, Nonthaburi, and Bangkok played the role of reference cities. These cases are illustrated in Table 5.10.

The Partners in Long-Term Urban Sustainability (PLUS) network[9] demonstrates another effective utilisation of a network by local governments in terms of city learning, as indicated by a survey conducted with cities participating in the PLUS network, as follows (Seymoar *et al.* 2009).

The survey findings show that learning from peers from other cities through intercity network programmes is recognised as one of the effective ways for local government officials to carry out daily tasks for sustainability. Face-to-face communication is considered to be particularly useful, in addition to web dialogue, newsletters, and information provided via websites, which are comparatively cheaper. Mutual learning among cities is taking place to solve the problems faced by local government officials.

Factors enabling local governments to adopt new practices via networks

Three international intercity environmental network programmes, ICLEI Southeast Asia, CITYNET, and the Kitakyushu Initiative, were selected for analysis of the processes and factors correlating to policies being adopted and diffused via networks.[10] This study involved an analysis of factors that corroborated policies

TABLE 5.10 State of intercity mutual reference (Kitakyushu Initiative)

Cities referring to other cities	*Cities referred to by other cities*
Kathmandu (Nepal)	Kitakyushu (Japan), Ulsan (Korea)
Dhaka (Bangladesh)	Kitakyushu (Japan), Nonthaburi (Thailand)
Bangkok (Thailand)	Kitakyushu (Japan), Surabaya (Indonesia)
Siem Reap (Cambodia)	Surabaya (Indonesia)
Nonthaburi (Thailand)	Kitakyushu (Japan), Surabaya (Indonesia)
Sibu (Malaysia)	Kitakyushu (Japan), Surabaya (Indonesia)
Bago (Philippines)	Surabaya (Indonesia), Bangkok (Thailand)
Surabaya (Indonesia)	Kitakyushu (Japan)
Weihai (China)	Ube (Japan), Surabaya (Indonesia), Beijing (China)
Dalian (China)	Kitakyushu (Japan)

Source: Nakamura *et al.* (2010b).

being adopted and disseminated via networks; in particular, it focused on the process of the adoption and diffusion of policies related to climate protection by ICLEI Southeast Asia, environmental education by CITYNET, and composting of organic waste by the Kitakyushu Initiative.

The results of the study imply that the factors noted below are important.

- First, with regard to internal factors in cities, the importance of support at the mayoral level was cited as an important factor.
- Next, the study found that stakeholder coordination and networking by individuals and organisations, both within and outside the local government, plays an important role.
- In addition, with regard to the autonomy of local governments from the central government and higher regional authorities, it was apparent that awareness of autonomy on the part of the city government, including the mayor, was important when looking at issues from the vantage point of jurisdiction, decision making, and financial affairs.
- At the same time, the study also determined that finance, another aspect of autonomy, was an important factor.

The support of the mayor, which is the primary promoting factor, is not always available for all cities, and some cities and mayors may not show any interest in participating in intercity collaborative activities. In such cases the attractiveness and relevance of practices dealt with in intercity collaboration would be key factors to motivate potential participants.

In looking at external factors, the study confirmed the following points.

- The existence of policy support from the national government and higher authorities, and financial support systems, particularly facilitated implementation of new policies.
- The study also showed examples that demonstrated the potential of financial and technological support from international development agencies and international non-governmental organisations (NGOs) in the implementation of new policies through active participation in international networks.
- However, with regard to expanding knowledge and experience from outside in terms of technological knowledge, intentional diffusion of technology does not occur when there is a lack of sufficient communication and discussion with local stakeholders, and insufficient tools and efforts to indigenise outside expert knowledge and experience.
- Adequate communication with stakeholders is also important when carrying out a thorough examination of issues and taking a multi-pronged approach to finding solutions.
- If cities do not introduce policies that contribute to solving local issues, even if policies are introduced, they will not be long-lasting without the support of stakeholders.

Cities in developing and in developed countries, and network secretariats and international development agencies as external stakeholders, should take note of these particular points when using and applying international networks.

Finally, the study findings included the following points in terms of the diffusion of practices via networks.

- Two factors – sharing a sense of purpose with similar cities through participation in networks, and disseminating information outside the city through networks – have the potential to contribute to improving committed support and the research capacity of stakeholders such as mayors and staff in related departments.
- The study also demonstrated cases where motivated participants intentionally expanded new policies to other cities.

Incentives and motivations for participation

According to a questionnaire survey[11] conducted on local governments participating in recent activities of the Kitakyushu Initiative, the following points were given as direct benefits of participating in the network:[12] (1) construction of facilities or implementation of technical cooperation projects with financial assistance from international organisations; (2) opportunities to build staff capacity through participation in study tours and seminars; and (3) expenses borne by the secretariat, which enables a city to participate in seminars and programmes. The following indirect benefits of participation in the network were cited: (4) increasing the recognition of the importance of environmental protection within the city, improving the position of environmental protection departments, and highlighting the importance of establishing departments in charge of the environment; (5) opportunities to solicit financial and technical support and intercity cooperation from overseas; and (6) sharing information within the city office about successful practices from other cities, for use in staff training and environmental education. Though there were differences in the perceptions of local governments about the direct and indirect benefits of participation in networks, economic benefits, such as in (1), (3), and (5), were cited first, and the effects of capacity building, such as is indicated in (2) and (6), were also pointed out. As shown in response (4), the potential to influence the negotiating capacity of environmental departments within local governments by participating in international networks was also expressed.

A study on the motivations for international environmental cooperation of Japanese municipal governments, which studied 12 Japanese cities,[13] revealed that these motivations can be divided into two types: materialistic and idealistic motivations (Nakamura *et al.* 2010a). The materialistic motivation emphasises the utilisation of local human capital for environmental management, promotion of environmental businesses, or prevention of trans-boundary pollution, while the idealistic motivation includes a sense of pride in sharing responsibility for environmental protection and conservation, or a sense of sharing global issues beyond national borders.

The cities of Kitakyushu and Yokohama are representative examples of the two patterns. Although socioeconomic factors such as exports, foreign residents, and foreign visitors, had been expected to facilitate cooperation, they were found to have limited correlation with it.

Previous and ongoing bilateral intercity collaboration in Asia: case study of Kitakyushu–Dalian, Kawasaki–Shenyang, and Yokohama–Penang

In this subsection, lessons learnt from city-to-city cooperation are examined, based on three cases of bilateral intercity collaboration: collaborations between (a) Kitakyushu and Dalian (China), (b) Kawasaki (Japan) and Shenyang (China), and (c) Yokohama and Penang (Fujikura *et al.* 2009).[14] The expected benefits and outputs of bilateral intercity collaboration are diverse. In the case of the Kitakyushu–Dalian cooperation, an urban environment management plan for Dalian City was developed and a Japanese official development assistance (ODA) loan was used for environmental investment projects in Dalian. In the case of the Kawasaki–Shenyang collaboration, international joint research has been developed in Shenyang based on research conducted in Kawasaki Eco Town, and Japanese recycling technology companies developed recycling businesses in Shenyang. The Shenyang City Government has also established a new administrative organisation for environmental model city development. In the case of the Yokohama–Penang collaboration, Penang developed a master plan incorporating Yokohama's experiences, which was followed by actual development of a town according to the master plan and a recycling pilot project.

The study also suggested that the commitment and vision of the mayors was important for continuous cooperation in the cities participating in bilateral intercity collaboration. One of the differences in multilateral intercity collaboration is the precondition of having a sister city relationship, as in the cases of Kitakyushu–Dalian and Kawasaki–Shenyang. The case of Yokohama–Penang also shows the potential for collaborative relationships developing from multilateral conferences, as Penang City attended an international conference hosted in Yokohama, showed an interest in Yokohama's urban planning, and requested the cooperation with Yokohama City (Tjandradewi *et al.* 2006).

Strategy to develop, operate, and utilise international intercity networks for sustainable development

Based on the discussions in the previous sections, this section discusses strategies for how to develop and best utilise international intercity networks for sustainable development, particularly from the perspective of local governments in Asia.[15]

Before referring to the above discussion, proposals for effective knowledge network management for sustainable development that were studied and proposed by the International Institute for Sustainable Development (IISD) will be discussed. The IISD performed experiments with formal knowledge networks to support change in specific policies and practices for sustainable development, and set out principles for managing knowledge networks for sustainable development (Creech and Willard 2001). Creech and Willard examine four stages in establishing a formal knowledge network in terms of strategies to seek effective management and governance of the network (see Table 5.11). Although an international platform for sustainable cities, or an intercity network, may not necessarily be the same thing as a knowledge network, the knowledge presented by the IISD is relevant and useful in considering strategies for developing and operating international intercity networks for sustainable development.

These ideas for steps from scoping to formalising a network would be applicable to international intercity network programmes. Local governments that are seeking the opportunities provided by an international network can start scanning and searching out potential partners first from various existing networks. Development of joint activities with specific partners may be achieved after participating in and utilising network programmes. Networking among local governments using an international network would include a range of activities from information provision and exchange, to nurturing relationships and trust, and to joint actions with specific partners. Activities differ according to the level of engagement of participating members and partners. The steps in starting to use an international network can be summarised as follows: (i) seek out existing networks; (ii) join planned/existing international networks; (iii) assess effectiveness in several trials; (iv) propose and create new programmes under the network; and (v) propose and lead the discussion and preparation of a new programme.

TABLE 5.11 Formulation of knowledge networks for sustainable development

Stages	*Actions*
1 Forming relationships	(a) Allocate time to seek out appropriate partners, (b) explore common interests that hold the network together, (c) learn realistic expectations of partner performance, (d) develop and apply criteria of membership, and (e) extend relationships beyond core membership, particularly to donors.
2 Organising relationships	(a) Develop goals, objectives, and work plans for the network, (b) decide how the network will be managed, and (c) explore options for financial resources particularly to support network coordination functions.
3 Formalising relationships	For members to discuss governance of the network: (a) begin with finalising agreement on goals, objectives, and principles, (b) codify the membership arrangement, (c) set decision-making rules regarding what requires consensus, majority votes, or merely input to the secretariat, and (d) customise the network structure.
4 Institutionalising relationships	(a) Be open to change processes and (b) review all relationships regularly to focus on solid, high-performing relationships.

Source: collated from Creech and Willard (2001: 58, 68, 79–81, 89–90, 92).

Another consideration in formulating an international intercity network for sustainable cities is whether it would be better to create a new one or to use an existing platform to initiate new programmes and activities. Developing a new platform requires a huge cost outlay, particularly for local governments. If national governments or international organisations make a commitment to secure necessary funds, there are still opportunities to formulate a new platform, but it might be better to consider initiating new programmes and activities under an existing framework. To try to extend the members of the new programme would be easier and more efficient for all stakeholders.

What kinds of benefits could a potential user of an international platform expect? As was shown above in the section "Incentives and motivations for participation" on the motivation of cities in developing countries and Japanese cities, the expected benefits include both tangible and intangible benefits. Tangible benefits for the cities in developing countries include economic gains such as project financing and organisational and individual capacity development for local governments. Tangible benefits for Japanese cities encompass promotion of local environmental businesses overseas, as demonstrated in the cases of Kitakyushu and Kawasaki. Intangible benefits for local governments in developing countries include access to information as well as funding opportunities, in addition to obtaining an international authority to help boost attention to environmental issues within the local government. Intangible benefits for the cities in developed countries would include promoting the city in the context of international competition, and a political demonstration of the city's environmental capital or of the city's contribution to global society. Such idealistic benefits might be considered to be an achievement in a city where there is political support in the constituency. In addition to local businesses, research institutions could benefit from collaboration with international intercity networks for sustainable development, as demonstrated in the case of the Shenyang–Kawasaki cooperation, in which research institutions in China and Japan developed collaborative relationships with local governments in both countries to extend their research and to contribute to policy formulation and assessment in China.

For an international intercity network for sustainable development, it is crucial to present the benefits of participation to mayors, local government officials, businesses, and researchers according to their respective priorities and interests, since the significance of potential benefits differ stakeholder-by-stakeholder. Mayors of cities in developing countries might be interested in foreign direct investment in the environmental business in their cities, while local government officials would be attracted by opportunities for their own capacity development. Alignment of interests and expectations among participating actors is a key factor in realising effective and functional international intercity networks for sustainable development. Different kinds of programmes could be designed and operated in an intercity network for sustainable development.

What, then, are the clues for how to realise and maximise the opportunities arising in an international intercity network for sustainable cities? Based on the intercity collaborations between Kitakyushu and Dalian, as well as Kawasaki and Shenyang, mutual benefits appear to be the key. In both cases, the city-to-city cooperation has targeted not only public-to-public technical (or financial) assistance but also business development, which ultimately seeks profitability. Strategic engagement of the business sector is one of the promotion factors to utilise in international intercity networks for sustainable development.

Another clue would be to link up committed and motivated cities that are eager to learn and provide knowledge and experience. The study on mutual reference and practice diffusion among three Asian intercity network programmes on the environment suggests that cities participate in network activities for capacity development more frequently if they are more eager to learn. Indeed, some of the cities that actively participated in the network activities actually adopted new practices such as organic waste composting and environmental education, as demonstrated by Surabaya, Nonthaburi, Phnom Penh, and Hanoi. Formulating a coalition of committed cities would enhance the potential of an international intercity network for sustainable development.

Governance, management, and platform operation are additional issues to be considered to achieve expected platform objectives. To respond to the actual needs and interests of participating cities, a platform needs to have effective communication and decision-making mechanisms that are trusted by members. If there are no channels to convey concerns and interests, and to properly reflect these in platform programmes and activities, participating cities lose motivation and a sense of ownership. The platform secretariat's capacity for internal communication and fundraising are also significant. Interaction with external and international actors, including international development agencies and banks as well as private companies, is expected of the secretariat of the platform, particularly for financial and institutional support and collaboration.

The benefits of utilising experts and international actors through an international platform are shown in the cases of city-to-city collaborations between Dalian–Kitakyushu, Shenyang–Kawasaki, and Penang–Yokohama, as well as in the UNEP Eco Town project. In the cases of cooperation between Dalian–Kitakyushu and Penang–Yokohama, local government officials provided the opportunities for peer-to-peer learning for the counterpart local governments. In the case of the Shenyang–Kawasaki collaboration, both local and international experts contributed to the Chinese city's policy development and implementation of a circular economy. The UNEP Eco Town project indicates that international experts have supported and will continue to support policy framework development as well as policy implementation regarding Eco Town, by providing advice and useful knowledge on relevant information and models.

However, difficulties were also observed in the above cases and studies. Notably, the lack of sufficient financing and a small number of committed staff for capacity development and practice diffusion were confirmed as problems in the studies (Nakamura *et al.* 2010b; Seymoar *et al.* 2009). The necessity of funding for cities in developing countries was also directly stated in one of the international platforms for sustainable cities: the sixth Asia-Pacific Eco-Business Forum in Kawasaki, Japan in February 2010. As well as showing that there are a limited number of cities with a strong commitment or motivation, the relatively smaller proportion of cities that frequently participate in network activities might indicate limited funding for inviting cities from developing countries. Insufficient funding might not be easily solved unless the network succeeds in raising funds from national governments, international organisations, private foundations, and so on. Yet the network could and should continue to develop attractive and meaningful activities within budget constraints to produce more motivated participating cities. This is an issue apart from identifying and linking cities that have already shown a strong level of interest.

What activities are possible for local governments participating in an international network? As described above, these would include knowledge and capacity development, fundraising and advocacy, business promotion, and joint implementation with regard to policies, procurement, research, and carbon financing. Apart from the existing cases in Asia examined in the previous section, there are several interesting activities being conducted in European and global networks. ICLEI Europe has been conducting a joint campaign called PROCURA+, which is promoting green procurement by local governments.[16] Products under this green procurement programme include buses, cleaning products and services, electricity, food and catering services, information technology products, and building construction. A planning and implementation guide comprising five milestones is provided, and manuals are available for participating local governments. The benefits for participating cities also include opportunities for seminars, conferences, and study tours, as well as discounted training and consultancy services.

Another example of international network activities at a global level is the purchasing consortium for clean technology products that has been organised by C40 Cities, an international intercity network to tackle climate change that involves 40 major cities, including London, New York, Tokyo, Hong Kong, Bangkok, and Jakarta.[17] C40 Cities is supported by the Clinton Climate Initiative (CCI), which is an initiative to combat climate change sponsored by the William J. Clinton Foundation. Pooling the buying power of participating cities helps to lower the price of energy saving products, and the CCI also introduces companies to cities, which has enabled more than 1,000 cities to access affordable energy efficient products.[18] These examples present future activities possible in an Asian platform for sustainable cities.

Last, here are some suggestions and requests for stakeholders in an international intercity network for sustainable development. For national governments, it is desirable to make collaborative agreements between governments to support intercity collaboration and to institutionally support intercity collaborative programmes. International organisations are also advised to link funding opportunities to local governments that utilise international intercity networks. The secretariats of existing intercity networks are expected to make it possible for a loose network to initiate new activities on specific issues under the auspices of external funders, to review performance and focus on performing activities to improve and strengthen the programme, and to explore collaboration among networks on specific topics by playing complementary roles in several networks. It would be also desirable to develop a brochure to disseminate the concept and examples of international intercity collaboration for sustainable development in various languages such as Japanese and Indonesian, to give local government officials the opportunity to understand intercity collaborations, and to consider participating in and utilising such collaborations.

International intercity platform for low carbon development

Applying an international intercity platform to the emerging issue of low carbon development would be desirable, given the achievements made so far in the field of traditional urban environmental management.[19] The issues facing an intercity network are what it would actually do and what its function should be. Possible ideas for Asian intercity collaboration on low carbon development are still being developed and there is room for further elaboration regarding the platform's functions and activities (Kitakyushu City 2010; Tjandradewi 2010). The Kitakyushu Initiative has already succeeded in diffusing the practice of composting organic waste (Maeda 2009), but it is important to generate more ideas for low carbon development policies and practices that could be diffused through intercity platforms. What kinds of activities are best done cooperatively, and could benefit from mediation and facilitation by a platform? This question needs to be addressed to realise the potential of intercity collaboration.

The secretariat of ICLEI Global has pointed out that there are several important sectors such as waste, transportation, and building over which local governments have some control that can effectively lower greenhouse gas (GHG) emissions at relatively low cost.[20] The Japanese Ministry of the Environment (MOE) has also identified the critical role of local governments in public engagement and urban design for low carbon development.[21] The functions and activities of international intercity collaboration need to be clarified to enhance the capacity and actions of local governments in these identified sectors.

The surrounding context in which cities in developing countries are situated must also be considered in an examination of potential international intercity collaboration on low carbon development. It is difficult for cities in Asian developing countries to implement policies that are solely aimed at building a low carbon society, because they have to prioritise more pressing development issues. Therefore, it is important to integrate climate change mitigation efforts into general development policies and measures. Several pioneering cities in Asian developing countries have initiated low carbon development policies and practices that also address additional problems at the same time. An international platform to support Asian cities in following low carbon pathways will need to address these problems.

Intercity collaborative activities for low carbon development

This section assesses four collaborative activities for low carbon development in terms of their potential to be the focus of intercity cooperation between Japanese cities and cities in developing Asia: (1) technical cooperation; (2) promotion of low carbon businesses; (3) carbon offsetting with engagement of local stakeholders such as citizens and companies; and (4) joint carbon credit projects of local governments. These four activity areas are derived from the typical scope of activities in international intercity collaboration for sustainable development, as explained above: mutual reference and learning, as well as capacity development; agreement and exchange for environmental business promotion; joint project implementation; and policy advocacy. Technical cooperation and business promotion correspond to the first and second categories. Carbon offsetting and joint carbon credit projects correspond to the third category (joint project implementation). No policy advocacy action is proposed here since the motivation to support such action does not seem to have developed so far in Asian local governments.

Technical cooperation

Technical cooperation, such as training of local government officials from developing countries in Japan and dispatching officials as experts to developing countries, has been a traditional focus of intercity cooperation. If the Japanese government provides institutional support to intercity technical cooperation on low carbon development, Japanese local governments can participate in technical cooperation with their existing knowledge and experience in the sectors of waste and wastewater treatment, air pollution, building and corporate energy efficiency, household energy consumption, and public transport and environmentally friendly vehicles. These areas also correspond to the current policies and practices of low carbon development by local governments in Asian developing countries. Such policies include incentive mechanisms, regulation, information provision, and education in each relevant sector, as well as action planning and organising and managing government units to promote low carbon practices.

Although Japanese local governments have contributed to JICA's technical cooperation in the environment field, such as waste management and recycling as well as air and water pollution control, cooperation specifically aiming at addressing climate change has just started with advanced local governments (see Table 5.12). This may be partly because local governments are still at the stage of trial and error in terms of formulating and implementing climate policy, and thus far they may not necessarily have attained clear, tangible outputs and achievements from these policies. Thus it is timely to establish a platform to facilitate and enhance technical cooperation by Japanese local governments in terms of low carbon development.

Tables 5.13 and 5.14 show commonalities and differences between Japanese municipal governments[22] and several cities in East Asian developing countries, including Dalian and Chongqing (China), Bangkok, Jakarta (Indonesia), and

TABLE 5.12 State of Japanese local governments' engagement in the JICA technical cooperation programme for the environment (2007–2009)

Field of cooperation	*Local governments engaged*	*Type of cooperation*
Climate change mitigation	Tokyo Metropolitan Government	Training in Japan
	Yokohama City	Training in Japan
	Nagoya City	Training in Japan
	Kyoto City	Training in Japan
	Kitakyushu City	Training in Japan
Water pollution	Sapporo City	Training in Japan
	Higashi-Hiroshima City	Training in Japan
	Minamata City	Training in Japan
	Osaka City	Training in Japan
Air pollution	Tokyo Metropolitan Government	Training in Japan
	Shizuoka Prefecture	Dispatching officials overseas
	Osaka Prefecture	Dispatching officials overseas
	Osaka City	Training in Japan
Waste management and recycling	Hiroshima Prefecture	Training in Japan
	Osaka City	Training in Japan Dispatching officials overseas
	Fukuoka City	Training in Japan Dispatching officials overseas
	Nagoya City	Training in Japan
	Kawasaki City	Dispatching officials overseas
	Kitakyushu City	Dispatching officials overseas
	Sapporo City	Dispatching officials overseas
	Kyoto City	Dispatching officials overseas
Urban environment management	Tokyo Metropolitan Government	Training in Japan
	Kanagawa Prefecture	Training in Japan
	Chiba Prefecture	Training in Japan
	Saitama Prefecture	Training in Japan
	Yokohama City	Training in Japan
	Kawasaki City	Training in Japan
	Chiba City	Training in Japan
	Saitama City	Training in Japan
	Sapporo City	Training in Japan
Natural environment conservation	Hokkaido Prefecture	Training in Japan
	Kushiro City	Training in Japan
	Obihiro City	Training in Japan

Source: JICA.

TABLE 5.13 Low carbon development policies and practices of Japanese municipal government as of 2008

Area of policy	*Policy tool*	*Contents*
Household	Financial/equipment assistance	• Financial assistance for photovoltaic (PV) household installations • Lending energy efficiency tools and data analysis services
	Facility and zone development	• Adoption of PV or solar heating in apartments • Collection of waste kitchen oil and utilisation for biodiesel fuel (BDF)
	Economic instrument	• Local-level carbon tax
	Awareness-raising and education	• Consultation service for energy efficiency • Children's programme on climate change and energy saving • Wind farm managed by or on behalf of local government • Woody biomass utilisation by local government • Demonstration of energy efficient houses • Capacity development of distributors and maintenance engineers of renewable energy equipment
	Institutional development	• Setting up organisations to enhance collaboration among local stakeholders • Collaborative activities with citizens for climate protection and energy saving
Industry	Financial/equipment assistance	• Assistance for biomass utilisation project
	Economic instruments	• Introduction of cap and trade system for carbon dioxide (CO_2) emissions • Carbon offsets • Taxation to mitigate climate change • Promotion of environmental finance and investment to mitigate climate change
Buildings	Regulations	• Introduction of environmental evaluation system regulations for buildings • Promotion and regulation of planting on the roof and walls of buildings
	Financial/equipment assistance	• Financial assistance for high energy efficiency houses • Grants for highly energy efficient equipment
	Facility and zone development	• Introduction of area-based heating and cooling and co-generation
	Economic instruments	• Taxation to mitigate climate change • Promotion of environmental financing and investment to mitigate climate change

continued

TABLE 5.13 Continued

Area of policy	*Policy tool*	*Contents*
Transport	Regulations	• Personal vehicle utilisation regulations • Parking regulations in central city areas • Introduction of transit malls • Introduction of car number regulation
	Financial/equipment assistance	• Grants for car pooling and bicycle use
	Facility and zone development	• Introduction of Light Railway Transit (LRT) • Construction of exclusive or preferential lanes for buses • Construction of bicycle lanes • Operation of community buses • Introduction of park and (bus) ride • Introduction of shared rental bicycles
	Economic instruments	• Introduction of road pricing • Taxation to mitigate climate change
	Institutional development	• Sharing distribution logistics

Source: Hosei University (2010).

TABLE 5.14 Low carbon development policies and practices of local governments in East Asian developing countries

Area of policy	*Policies and practices*
Climate change planning	• Action plan development for climate change mitigation • Setting targets for GHG emissions reduction by the city
Household	• Biomass utilisation by farmers • Reduction of municipal waste • Composting of organic food market and community waste • Fuel switching from kerosene to liquid petroleum gas • Environmental education, awareness-raising, and plain language information provision for citizens
Industry	• Corporate energy efficiency programme • Corporate energy audit • Eco-industrial parks for resource circulation and recycling
Building	• Eco model housing
Transport	• Extension of public transport such as bus rapid transit and subway • Fuel switching • Street lighting retrofitting
Urban greening	• Planting • Urban park development and street planting

Source: papers presented by each local government at the international workshop on local initiatives towards low carbon development in Asia, on 24 February 2010.

Ho Chi Minh City (Vietnam) in terms of their policies and practices for low carbon development. Policies such as overall planning for climate change, energy efficiency improvement programmes in the industrial sector including small and medium-sized enterprises (SMEs), and raising awareness in and providing information to the household sector are common to local governments both in Japan and in Asian developing countries. In contrast, Japanese local governments do not have experience in small-scale biomass methane fermentation to supply energy or bus rapid transit, for instance. The common policy and practice areas are primary candidate fields for technical cooperation in a platform to match Japanese experience and local relevance in Asian developing countries.

Local governments in Asian developing countries may need to consider adopting systems to measure and report their GHG emissions, and Japanese local governments may be in a good position to provide related technical assistance.[23] Japanese local governments have been accumulating experience in this area since 1999, when they were required to introduce such systems. In addition, Japanese local governments could share their experiences on GHG emissions estimation and monitoring once they become familiar with global measurement and reporting standards such as the International Local Government GHG Emissions Analysis Protocol, drafted by ICLEI in 2009, and the Public Sector Accounting Reporting Standard, under preparation by World Resources Institute in 2010 (Hosei University 2010).

It should be noted, however, that the number of Japanese local governments that are interested in international collaboration on low carbon development is not large. Among 13 Eco Model Cities, which were selected out of 89 cities and municipalities by the national government in 2008 and 2009 as cities that have initiated low carbon policies and practices and have set ambitious goals and strategies to realise a low carbon city, there are four cities that have explicitly referred to international collaboration in their strategies, namely Kitakyushu, Yokohama, Kyoto, and Minamata. Among other cities that have applied for Eco Model City status and have not yet been selected, only Kawasaki and Hiroshima mentioned international dissemination of their experiences in their proposals.

Japanese local governments will find it increasingly difficult to implement international technical cooperation solely dependent on their own budgets and human capacity. Although peer-to-peer learning among local governments would be effective for knowledge exchange, capacity development, and practice diffusion to a certain degree,[24] financial and personnel assistance are needed to maintain and promote such local government initiatives.[25] Since Japanese local governments are under strong pressure to reduce fiscal expenditures, international cooperation activities are currently given a low priority even in cities that previously participated in international cooperation. For Japanese local governments international collaboration is not easy because of personnel and fiscal constraints.[26] Lack of human capacity and experience in directly communicating and coordinating with foreign organisations is also viewed as an obstacle.[27] International collaboration could be possible for these cities when external organisations in Japan assist them in terms of communication and coordination with foreign partners.

It is therefore desirable for there to be a platform to address these obstacles experienced by Japanese local governments as much as possible. Such a platform could mediate between local governments in Japan and in Asian developing countries, and develop training programmes on technical cooperation by local governments on low carbon city development.

The Japanese government and national-level organisations such as JICA should be encouraged to authorise international cooperation for low carbon development in Asia as an aspect of foreign policy. They should provide sufficient policy commitment and institutional support for implementation to enable Japanese local governments to collaborate with local governments in Asian developing countries, for example by assisting with a cooperation platform.

In collaboration with the platform secretariat, JICA could develop a training programme for Asian local governments that focuses on capacity development for low carbon development planning and management. This could include core training at a JICA regional centre and field visits to relevant Japanese local governments so that trainees from Asian developing countries could assess the current situation and identify possible plans and programmes. To promote south–south cooperation[28] among cities, a scheme of training in a third country would be effective (i.e. training programmes conducted outside Japan, in a developing country), where trainees are invited from several developing countries, with the support of the platform secretariat.

When the concrete scope of technical cooperation on low carbon development policies and practices is discussed and defined by relevant local governments and development agencies, this should also provide an account of integrating the needs of adapting to climate change, particularly in the area of urban planning and regulations (JICA 2007).

Promotion of low carbon business in Asia

Japanese local governments of areas that have a concentration of environmental and energy saving businesses that could contribute to low carbon development in their jurisdiction would have the potential to promote international business collaboration with Asian cities, such as those in China, to support Asian business development of local SMEs, which are confronting saturation in domestic markets. Examples of low carbon businesses relevant to local government mandates include energy service companies (ESCOs) in the building and corporate sectors, energy efficient electrical appliances in the household sector, and perhaps environmentally friendly vehicles in the transport sector. Business promotion was the intention of Kitakyushu City Government in launching its Kitakyushu Asian Center for Low Carbon Society. International collaboration, in this case, is not easy for Japanese local governments without active policy and institutional support from the national government and organisations such as the Industrial Cluster Program of METI, and various supporting programmes of the Japan External Trade Organization (JETRO).[29]

There are three regions in Japan that could have the potential to further extend local governments' support to low carbon business promotion in Asia, based on the national-government led industrial policy to develop regional industrial clusters that are relevant to low carbon development,[30] namely the Kyushu, Kinki (Greater Osaka), and Kanto (Greater Tokyo) regions. Although the Regional Bureaus of the Ministry of Economy, Trade and Industry (METI) and JETRO have extended their support to local governments interested in international environmental business promotion in Asia, there are a limited number of Japanese companies, particularly SMEs, which are interested in overseas business development with governmental support, and local governments are still faced with financial limitations. Results have thus been limited so far, and a more effective and efficient way of environmental business promotion is needed.

An international platform on low carbon cities could support collaboration among neighbouring cities and between different regions in Japan in order to enhance business opportunities. Strategic alliances within and between regions in Japan[31] could be another option, which has not been tried yet, for relevant local governments to overcome resource limitations in terms of financing and the number of appropriate companies in their locality. By using an intercity platform, collective operating cost reduction by pooling resources can be explored. Increased numbers of participating cities and companies and increased access to potential markets through a platform would increase the chance of successful business matching. The platform secretariat should also be encouraged to develop supporting programmes that are appropriate for local governments, such as international regional agreements and coordination between economy and environment bureaus within city governments.

At the same time, technical cooperation on environmental regulations and institutions is also relevant to business promotion. In business areas where regulations are necessary for certain technologies to be adopted, local governments with environmental industry assets should consider technical cooperation on environmental regulations and institutions and select target countries and cities with the aim of future business development in Asian countries. Examples of such technologies include cleaner production, recycling, and pollution treatment. Without regulations mandating recycling or pollution reduction, companies do not have strong incentives to adopt new technologies even if they are theoretically good investments that reduce operational costs, in addition to the environmental improvement effect.

Promoting low carbon business in Asia would be a promising activity for an international low carbon city platform. Assistance from the national government, however, is still needed because of the current limited and scattered resources. It is desirable for Japanese local governments to use national-level support and institutions provided by METI, JETRO, and others as they have done thus far. Business collaboration could eventually contribute to revitalisation of local communities in Japan through increasing exports to and direct investments in Asian developing countries. The expected support functions of the platform secretariat include needs and seeds surveys, mediation between local governments, and matchmaking between businesses and local governments. The platform could also provide a venue for coordination of various relevant partners and provide necessary information, knowledge, and services.

Joint carbon credit projects

Japanese local governments and local governments in Asian developing countries could jointly develop projects that produce carbon credits to certify GHG emissions reduction, where Japanese local governments would purchase the carbon credits produced. This could be considered once Japanese local governments are mandated to reduce or cap their GHG emissions from the businesses and projects they directly manage. Even in that case, they might have the preference to purchase carbon credits on the market, or to implement joint projects only in Japan. Careful assessment of cost-effectiveness as well as political support from local stakeholders in Japanese cities will be required when they examine the option of formulating joint projects.

Carbon offsetting with citizens' and corporate engagement

Although it is not necessary to develop international linkages between cities to encourage citizens to engage in low carbon development, carbon offsetting is a possible way to engage citizens with international intercity collaboration on low carbon development in order to promote environmental education on climate change mitigation. Japanese local governments could offset carbon emissions from local events with the participation and support of citizens and local enterprise, using voluntary carbon credits produced from GHG emissions mitigation projects in partner cities in Asian developing countries, which operate with the support of a Japanese city. This brings about another opportunity for awareness raising, environmental education, and perhaps international development education for Japanese cities. In addition, it may also be possible for local companies to offset part of their GHG emissions using the credits provided by such intercity relationships between local governments as a corporate social responsibility (CSR) action, since the Japanese voluntary carbon market has grown steadily, although it is still not very large compared to the regulatory market. At this moment, it would be difficult for Japanese local governments to cooperate with projects in cities in Asian developing countries to implement carbon offsetting or eco point utilisation because of the difficulties in securing initial investment and budgets for operating these projects.[32] Japanese local governments consider that it is inappropriate to use offsetting when the validity of carbon credits is not guaranteed.[33] However, it might be feasible to use eco points to support low carbon community development projects in Asia, since it can be treated as a contribution that emphasises developmental effects, not as carbon offsetting for reduction of GHG emissions. In this case, the ambiguity in terms of GHG emissions reduction may not harm the essence of the collaboration.

It would be better to use such a contribution to projects that do not provide validated carbon credits as an environmental education tool, particularly to promote Japanese citizens' understanding of global aspects of climate change issues.

As part of environmental education and international development education, it might be feasible for Japanese local governments to cooperate with GHG emissions reduction/absorption projects in sister cities in Asian developing counties when there is no clear declaration of obtaining carbon credits or offsetting as a benefit of cooperation. This would give citizens and local students some understanding of the global nature of climate change mitigation and the different conditions in developing countries.

Several Japanese local governments have initiated domestic inter-regional collaboration to use carbon offsetting, including regional agreements between (a) Shinjuku Ward, Tokyo and Ina City, Nagano Prefecture, (b) the Tokyo Metropolitan Government, Chiyoda Ward, Tokyo and Aomori Prefecture, and (c) Yokohama City, Yamanashi Prefecture, and Doshi Village, Yamanashi. Here, a local government in an urban area supports a counterpart in a rural area, either financially or in terms of labour, regarding forest management or renewable power generation, with a certificate or declaration of contribution to GHG absorption. These domestic collaborative activities could provide the basis of operations for extending such collaborations to local governments overseas. Yet it would be difficult for Japanese local governments to communicate with their counterparts in Asia without language barriers, and difficult to monitor a project overseas. An international platform might be able to mediate between them and provide monitoring and reporting services on behalf of counterparts in Asian developing countries.

Another theoretically possible way to engage citizens in international intercity collaboration on low carbon development is utilisation of eco points. Japan uses an eco point system to reward Japanese citizens for adopting low carbon practices such as increasing their use of public transport. Eco points could also be used to fund low carbon projects in partner cities in Asian developing countries. There are some difficulties in incorporating international collaboration into eco point projects that are conducted in Japanese cities because of (a) the difficulties in securing initial funds that are observed in eco point projects, (b) difficulties of fiscal autonomy, and (c) the small amount of initial funds available compared to the amount of funds needed to assist a GHG emissions reduction project of a meaningful size in Asian developing countries. However, once the necessary funds are secured along with a sufficient number of participating Japanese citizens, a platform could provide mediation and monitoring services.

Roles of a platform secretariat to facilitate collaboration between local governments

The existence of an effective platform secretariat is critical in terms of mediation between cities, identification of low carbon development programmes in developing countries, development of training programmes and materials to ease Japanese local governments' technical cooperation, GHG emissions measurement capacity development to visualise the effects of international cooperation, and extending good practices by creating linkages with donors' financial assistance to development programmes in developing countries. The Japanese government and other potential donors could contribute to this institutional setup for local-level low carbon development in Asia.

Particularly for technical cooperation between local governments, in order to incorporate the different interests of developing countries and developed countries in terms of GHG emissions reduction, it is desirable for a secretariat to create linkages between identification and formulation of co-benefit type low carbon development programmes, measurement of GHG emissions reduction, and financial and technical support. Of these activities, identifying and formulating low carbon development programmes in developing countries and measuring and estimating GHG emissions reductions can be conducted within the capacity building for a clean development mechanism (CDM) implemented by JICA and IGES, which are supported by the Japanese national government. The point here is that it is possible to utilise such existing capacity-building schemes in order to support an intercity platform for low carbon development, focusing on low carbon development programmes that are executed or coordinated by local governments in Asian developing countries. Knowledge production from Japanese experience and knowledge sharing with other countries, and coordination across national borders, could also be included in the secretariat's role. The secretariat could develop strategic partnerships with local environment research institutes connected to Japanese local governments in order to produce and compile the relevant knowledge and experience.[34]

Regarding low carbon business promotion, the secretariat could conduct needs and seeds surveys, mediate between local governments, and provide opportunities for matchmaking between businesses and local governments that are interested in low carbon business promotion.

Funding of the secretariat has been a problem for most intercity network programmes for the environment. To realise the above-mentioned roles of a secretariat, it would not be possible for member cities to bear all the necessary costs or to provide the required human resources through membership fees or in-kind contributions. The support of national governments in developed countries, international financial development agencies, or private foundations is needed in this regard. The next subsection touches on the possibility of the Japanese government's support.

Role of the Japanese government

Considering the limited resources of Japanese local governments with regard to international environmental cooperation, the Japanese national government should provide institutional support to make it easier for local governments to contribute to national policy, focusing on support for an international platform that could host integrative programmes for intercity collaboration on low carbon development, in addition to the existing project-based support. Although Japanese local governments have experience with low carbon technology and policy in some areas, such as waste and wastewater management and street lighting energy efficiency, further collaboration with third parties is needed to introduce simplified measurement and reporting of GHG emissions, especially if Japanese local governments are to extend technical cooperation to local governments in Asian countries.

Conclusion

Based on the above preliminary assessment of possible typologies of intercity collaboration on low carbon development, which are summarised in Table 5.15, there are two major possibilities for the focus of functions and activities of an international platform for low carbon cities in Asia: technical cooperation and business promotion. Figure 5.2 shows a schematic representation of possible intercity collaborative activities and the associated roles of an international platform on low carbon cities.

One option would be technical cooperation on low carbon policies and practices between Japanese local governments and local governments in Asian developing countries (Figure 5.2). Participating local governments would be expected to exchange and share policies and practices via the international platform on low carbon cities. The Japanese national government would be requested to provide

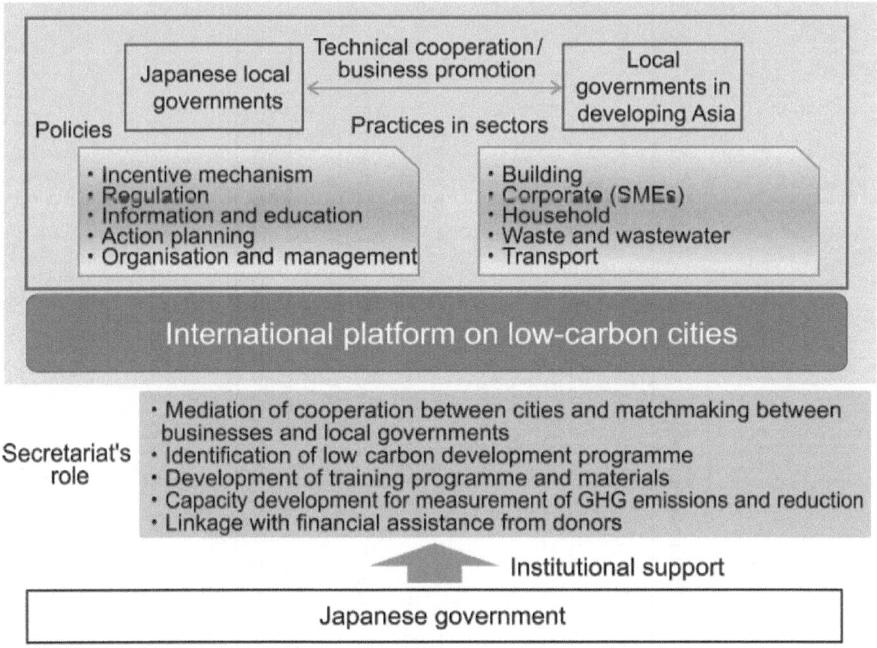

FIGURE 5.2 Concept of a platform for international intercity low carbon collaboration in Asia (source: Nakamura 2010a, with minor amendment).

TABLE 5.15 Preliminary assessment of intercity collaboration on low carbon development in Asia

Types of collaboration	*Benefits to Japanese cities*	*Benefits to cities in Asian developing countries*	*Current supportive policy*	*Needed future policy*	*Barriers to implementation*
Technical cooperation	Promoting the city as environmentally progressive	Capacity development	JICA's scheme	Support focused on local government	Low priority, financial constraints
Promotion of low carbon business	New business opportunities	New business opportunities	METI/JETRO's support	Support for domestic alliances and coordination in Japan	Scattered resources, insufficient financing
Joint carbon credit project	Meet GHG emissions reduction targets	Local development	None	–	No incentive without GHG emissions caps
Carbon offsetting with citizens' engagement	Environment/ development education	Contribution to local development	JICA's scheme	–	Minimal GHG emissions mitigation

Source: Nakamura (2010a).

assistance so that the platform secretariat could provide necessary and effective support to participating cities, including mediation between cities, identifying low carbon development programmes, developing training programmes and materials, capacity development for GHG emissions measurement, and linking with donors' financial assistance.

Another option would be collaborative agreements on low carbon technology business promotion between Japanese local governments and local governments in Asian developing countries with the collaboration of businesses that have, or are willing to have, low carbon technologies. The Japanese national government as well as business associations could assist the platform secretariat so that it could provide needs and seeds surveys, mediation between local governments, and match-making between businesses and local governments.

Both cases suggest that the Japanese government should provide institutional support to local governments with regard to financial and personnel resources, including development of a special budget account for international collaboration by local governments. Establishment of a designated office or organisation for the international platform would be desirable.

The Kitakyushu Asian Center for Low Carbon Society and Asian City Network for Environment Improvement could be a good candidate for a platform to perform the functions mentioned above. It will soon begin operations, but what projects and activities it will undertake are still under consideration.

Cooperation among platforms is also desirable. For instance, to extend networks among Japanese local governments, CITYNET could collaborate with Japan's Promotion Council for Low Carbon Cities (PCLCC), which includes 85 cities, 46 prefectures, 12 ministries, and 25 governmental organisations, and which is chaired by the Kitakyushu City Government.[35] To gain greater access to Asian cities in developing countries, it is recommended that the PCLCC collaborate with CITYNET, which has 73 member cities in developing Asia. In terms of financing new programmes, the Japanese government could support the platform secretariat both for technical cooperation and business promotion, while relevant Japanese private companies and business associations could support it both financially and with in-kind contributions. Collaboration between different platforms with different resources could cost less than creating a new platform, and enable optimum utilisation of complementary assets and capacities.

To examine the feasibility of the above-mentioned international platform and its potential collaborative activities, further study and close consultation with potential stakeholders and partners are needed.

Notes

1 This chapter was supported by the Environmental Research and Technology Development Fund (B-0701, E-0906) of the Ministry of the Environment, Japan.

2 Chair's Summary, "High Level Seminar on Environmentally Sustainable Cities under the Framework of the East Asia Summit Environment Ministers Meeting". See www.hls-esc.org/1st_HLS_ESC/HLS_ChairsSummary_final.pdf.

3 This subsection is from Nakamura *et al.* (2010b).
4 The websites of these networks are as follows: Kitakyushu Initiative, www.iges.or.jp/kitakyushu/; CAI-Asia, www.cleanairnet.org/caiasia/1412/channel.html; ICLEI, www.iclei.org/index.php?id=586; and CITYNET, www.citynet-ap.org/En/user/home/home.php.
5 ICLEI is a worldwide network, but for the present purpose the focus is on ICLEI Southeast Asia, which is part of the ICLEI global network.
6 From the websites of each network programme.
7 This subsection is based on Nakamura *et al.* (2010b).
8 For example, 12 cities attended CITYNET network events at least 11 times during the 22 years of the network's operations.
9 The PLUS network is "a network of cities, towns and regions sharing their experiences, expertise, and tools to undertake urban sustainability plans and projects", and has been renamed the Sustainable Cities International Network. The secretariat is Sustainable Cities International, located in Vancouver, Canada. See http://sustainablecities.net/citiesnetwork.
10 This subsection is from Fujikura *et al.* (2009) and Nakamura (2010b).
11 The questionnaire was sent out to 17 cities and responses were received from eight.
12 This subsection is based on Nakamura *et al.* (2010b).
13 Kitakyushu, Kawasaki, Yokkaichi, Ube, Fukuoka, Minamata, Osaka, Niigata, Kushiro, Yokohama, Kyoto, and Nagoya.
14 This subsection is based on Fujikura *et al.* (2010).
15 This section is mainly based on Fujikura *et al.* (2010).
16 www.procuraplus.org/.
17 www.c40cities.org/.
18 www.clintonfoundation.org/facts/content/william-j-clinton-foundation-overview.
19 This section is from Nakamura (2010a).
20 Presentation at an ICLEI Japan seminar on the path towards a low carbon society, on 15 February 2010, in Tokyo, Japan.
21 Discussion during the above seminar.
22 From Hosei University (2010); 1,111 municipal governments responded to the survey. "Municipal government" includes cities, wards, towns, and villages in the Japanese administration system. Prefectural governments, intermediate local governments between national and municipal government in Japan, were not included in the study.
23 Based on the experiences of ICLEI Cities for Climate Protection in Indonesia and of C40/Clinton Climate Initiative's Degree 2 project in Indonesia.
24 For example: Tjandradewi and Marcotullio 2009; Tjandradewi *et al.* 2006; Seymoar *et al.* 2009; Nakamura *et al.* 2010a.
25 Nakamura *et al.* 2010a.
26 Interview with Kanagawa Prefecture, Environment and Agriculture Department, Environmental Planning Division, on 27 October 2009.
27 Ibid.
28 South–south cooperation refers to international cooperation among developing countries, which is comparable to north–south cooperation between developed and developing countries.
29 Kitakyushu City has been conducting Japan–China sound material cycle city cooperation projects, or Eco Town projects, with the Chinese cities Qingdao, Tianjin, and Dalian, under the framework and support of METI, since FY2007, 2008, and 2009, respectively. A company in Kitakyushu City has been conducting a demonstration project in Dalian, China, for plant facility energy conservation since 2008 using a JETRO scheme. Concrete collaboration ideas between Kitakyushu Asian Center for Low Carbon Society and METI from 2010 are under consideration (interview with Kitakyushu City Environmental Bureau Environmental International Cooperation Division, on 25 March 2010).
30 METI's regional industrial clusters can include various business fields other than the

environment. Kyushu, Kinki, and Kanto are the three regions that have clusters relevant to low carbon development. METI has nine regional bureaus in Japan. See www.cluster.gr.jp/en/index.html.

31 Though Japanese cities may see themselves in competition with each other in environmental business, an alliance makes sense because there are still few SMEs interested in overseas business development (interview with Kyushu Recycle and Environmental Industry Plaza, on 4 December 2009).

32 Eco points are a mechanism to promote individual environmentally friendly actions, and are similar to local currency, or eco money. In this mechanism, individuals obtain eco points through environmentally friendly actions such as refusing the use of plastic bags for shopping, and then can later use the points collected to gain benefits such as local gift certificates. Benefits are usually provided by local businesses and/or local governments that participate in the eco point programme.

33 Interview with Kitakyushu City Government, Hiroshima City Government, and Kawasaki City Government on 17 August 2009, 2 December 2009, and 16 October 2009, respectively.

34 This would also contribute to redefining and revitalizing local environmental research institutes, which has been an issue in itself. Kawasaki City started to revitalize its environmental research institute by adding a new mandate of international cooperation through information provision.

35 www.kantei.go.jp/jp/singi/tiiki/kankyo/soukai_dai2/2siryo.pdf.

References

CAI-Asia and CDIA (2009) *City networks survey and roundtable report: Network of city networks project*, Manila: CAI-Asia and CDIA.

Creech, H. and Willard, T. (2001) *Strategic intentions: managing knowledge networks for sustainable development*, Manitoba: International Institute for Sustainable Development.

Fujikura, R., Mori, H., Sano, D., and Nakamura, H. (2009) *Higashi ajia de no jizoku kano na toshi seisaku gijutsu shinario no prattofomu [Platform on policy and technology scenario for sustainable cities in East Asia]*, Hayama, Japan: Institute for Global Environmental Strategies (IGES).

Fujikura, R., Mori, H., Sano, D., and Nakamura, H. (2010) *Draft guideline of international platform for sustainable cities*, Hayama, Japan: IGES.

Hosei University (2010) *Nihon no jichitai ni okeru teitanso shakai kouchiku oyobi chikyu kankyo mondai heno torikumi sokushin shisaku ni kansuru kenkyu [Study on policies of low carbon society development and global environment problems in Japanese municipal governments]*, Tokyo: Hosei University.

IGES (2010) *Key documents on Clean Asia Initiative (CAI) promotion: October 2009 – March 2010*, Hayama: IGES.

JICA (Japan International Cooperation Agency) (2007) *Kikohendo heno tekiosaku ni kansuru JICA no kyoryoku no arikata [Methods of JICA cooperation for policies on adaptation to climate change]*, Tokyo: JICA.

Kitakyushu City (2010) "An Asian base for green (low carbon) revolution: Kitakyushu Asian Center for Low carbon Society", Paper presented at international workshop on local initiative towards low carbon city in Asia, Fukuoka, Japan, 24 February.

Maeda, T. (2009) *Reducing waste through the promotion of composting and active involvement of various stakeholders: replicating Surabaya's solid waste management model*, Hayama, Japan: IGES.

Nakamura, H. (2010a) *Enhancing low carbon development through international cooperation between cities in Japan and in Asian developing countries: roles and activities for an international platform on low carbon cities in Asia*, Hayama, Japan: IGES.

Nakamura, H. (2010b) *Political factors facilitating practice adoption through Asian intercity network programmes for the environment*, Hayama, Japan: IGES.

Nakamura, H., Elder, M., and Mori, H. (2010a) *Explaining international environmental cooperation by Japanese municipal governments with developing countries*, Hayama, Japan: IGES.

Nakamura, H., Mori, H., and Elder, M. (2010b) *Mutual learning through Asian intercity network programmes for the environment*, Hayama, Japan: IGES.

Seymoar, N.-K., Mullard, A., and Winstanley, M. (2009) *City-to-city learning*, Vancouver: Sustainable Cities.

Tjandradewi, B.I. (2010) "Low carbon cities in Asia and CITYNET", Paper presented at international workshop on local initiative towards low carbon city in Asia, Fukuoka, Japan, 24 February.

Tjandradewi, B.I. and Marcotullio, J.P. (2009) "City-to-city networks: Asian perspectives on key elements and areas for success", *Habitat International*, 33:165–72.

Tjandradewi, B.I., Marcotullio J.P. and Kidokoro, T. (2006) "Evaluating city-to-city cooperation: A case study of the Penang and Yokohama experience", *Habitat International*, 30:357–76.

Yano Research Institute (2009) *Kyuseicho suru onshitsu koka gasu haishuturyo bijinesu no jittai to tenbo* [*State of and perspectives on the rapidly-growing business of greenhouse gas emissions trading*]. Tokyo: Yano Research Institute.

PART II

Cases of climate change mitigation

6

SECTORAL POLICIES FOR GHG MITIGATION IN ASIA

Policies for energy, industry, transportation, and buildings in China and India

Frank Hiroshi Ling

Introduction

The Asia-Pacific region, home to 60 percent, or nearly four billion people, of the world's population, is undergoing rapid urbanization and industrialization. While Asia accounts for 35 percent of global greenhouse gas (GHG) emissions from energy consumption today, that figure may reach up to 45 percent by 2030 (IEA, 2011) with more than 60 percent of new global GHG emissions coming from this region each year. China and India, which account for populations of 1.3 billion and 1.1 billion, respectively, are the main drivers of this trend. The use of fossil fuels, which releases the main GHG, CO_2, is the only affordable option for meeting their growing energy needs in cities, industries, and transportation systems. Figure 6.1 shows that China and India account for large share of the new primary energy demand and fuel growth and their associated emissions.

The accumulation of atmospheric CO_2 and other GHGs underlies the current trajectory in climate change. There is overwhelming scientific consensus that levels of atmospheric CO_2 should not exceed more than twice the pre-industrial concentration of 280 ppm in order to prevent irreversible changes to the world's climate. As a result of human induced climate change, the average global temperature went up by 0.74°C between 1905 and 2005. If the current trends in consuming energy continue, this will lead to devastating effects on ecosystems, industries, agriculture, and transportation systems around the world.

It is estimated that stabilization of CO_2 levels at roughly 500 ppm could be achieved by holding emissions at the 2008 level of eight billion t-C per year over the next few decades, followed by drastic cuts of up to 80 percent by 2050 (Pacala and Socolow, 2004).

Technologies will play a crucial role in achieving mitigation targets by lowering both energy and carbon intensities. The Intergovernmental Panel on Climate

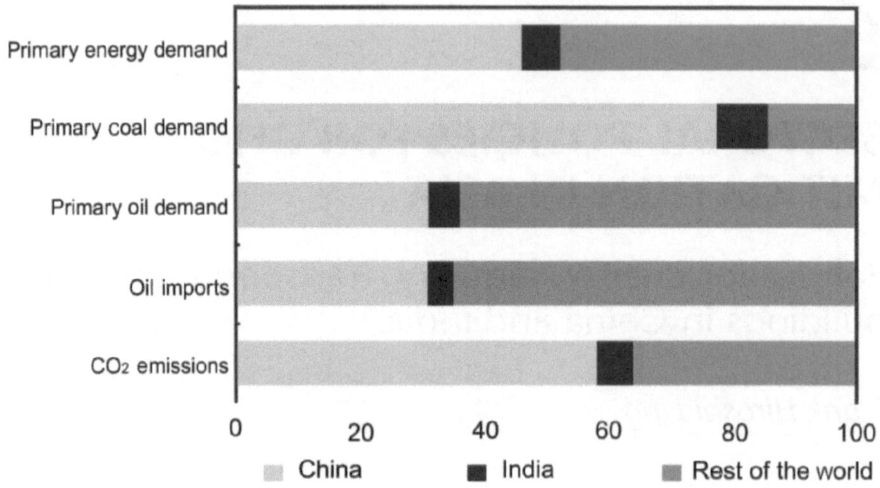

FIGURE 6.1 Share of incremental energy demand, imports and energy related CO_2 emissions growth between 2000 and 2006 (source: adapted from IEA, 2007b).

Change (IPCC) has stated that technologies already exist to stabilize and reduce GHG emissions (IPCC, 2001). Energy efficiency is arguably the most readily accessible technology for lowering carbon intensity in industries, buildings, and power generation. In particular, there is a broad set of technologies already available for enhancing efficiency of lighting, heating, cooling, refrigeration, and other appliances. In the longer term, future wide-scale deployment of renewable energy systems will replace the current energy infrastructure with one that is not reliant on fossil fuels. These renewable technologies include solar, wind, geothermal, hydro, and biomass power. Yet, technology alone will not be sufficient to meet climate goals. The solution to achieving effective mitigation lies in an integrated approach that takes into account developmental needs, aspirational goals, cultural values, political systems, and socioeconomics.

In Japan, researchers have led efforts to develop integrated assessment tools to help decision makers create viable roadmaps to transition Asian economies toward a low carbon society (LCS) and to evaluate various policy options for GHG emissions reductions (Fujino *et al.*, 2008). For developing countries, the key to mitigation is to avoid high carbon energy development. In contrast to industrialized countries that are characterized by patterns of high energy and carbon intensities, Asian countries can avoid a technology "lock-in" by circumventing conventional patterns of development. Enhancing energy access while restraining the growth of GHG emissions is, therefore, a great challenge for the region.

Due to their immense size and populations, Asian countries, namely China and India, represent perhaps the greatest challenge and opportunity for developing and deploying technologies to reduce carbon emissions. They have low per-capita emissions today compared to industrialized nations, and their aspirations are driven by a desire to reduce poverty and improve their standards of living. Each year, approximately 44 million people migrate from rural regions to urbanized environments in Asia (ADB, 2007a), causing a sharp increase in energy use and motorization and energy intensive industries.

Both China and India have been taking substantial steps toward a low carbon future in the context of their developmental needs. In this chapter, we review how two of the most populous countries in the world are promoting climate mitigation in major sectors of their economy: energy production, industry, buildings, and transportation. In addition, the latest initiatives in low carbon urban development will be discussed.

China

A look back at recent Chinese history shows the country's potential for substantial reductions in current and future GHG emissions. Between 1980 and 2000 as China underwent reforms and transitioned to a market driven economy, it dramatically improved its energy intensity. As the country quadrupled its GDP, it not only lifted hundreds of millions out of poverty but at only a doubling of its overall energy use. However, carbon intensity grew between 2002 and 2005 as China's economy continued to grow at breakneck speed. By 2006, more growth in emissions was seen over the past four years than in the previous quarter century, and, in 2009 China became the world's largest producer of GHG emissions. Yet, there are abundant opportunities for GHG emissions abatement, and preliminary studies indicate that China, the source of 80 percent of Asia's industrial growth over the past 25 years (IEA, 2007a), may also have the world's largest technical emission reduction potential of approximately 3.5 billion t-CO_2 by 2020 (Hanaoka *et al.*, 2008).

The central government of China has undertaken major steps to reduce energy intensity. China's 11th Five Year Plan (FYP) for the period between 2006 and 2010 called for an ambitious 20 percent target improvement in energy intensity per unit of GDP from 2005 levels by 2010. A preliminary analysis shows that the target has been met through national programs including the Ten Key Projects, Buildings Energy Efficiency, the Top-1000 Program, Appliance Standards and the Small Plant Closure Program (Price *et al.*, 2011). Table 6.1 shows an estimate of the reduction in energy consumption and emissions between 2005 and 2008 under these policies.

In addition, the plan called for the reduction of major pollutants by 10 percent from 2005 levels, including CO_2 and SO_2. The target was based on the country's 2004 National Development and Reform Commission's (NDRC) mid-term energy conservation plan that aimed to reach 1990 international energy efficiency levels by 2010 and catch up with international levels by 2020. The National Climate Change Action Program (NCCAP) also called for a 20 percent increase in forest cover over China's land mass by the end of 2010. In addition to the Five-Year Plan, the Chinese government has also instituted the NCCAP (2007) to outline specific targets to address climate change through energy efficiency, renewable energy, nuclear power, land use and forestry, and technology development (Lewis, 2007).

TABLE 6.1 Savings in primary energy consumption and emissions in China between 2005 and 2008

	Primary energy (Exajoule: EJ)	*Emissions (million t-CO_2)*
Ten Key Projects	3.0	287
Buildings energy efficiency	1.2	100
(Overlap – Ten Key Projects and buildings energy efficiency)	–1.2	–100
Top-1000 program	3.6	197
(Overlap – Ten Key Projects and Top-1000 program)	–0.4	–27
Small plant closures	2.7	171
Appliance standards	1.1	78
Other savings including provincial programs	5.4	612
Total primary energy savings	15.4	1,318

Source: adapted from Price *et al.*, 2011.

Today, Chinese leaders recognize the devastating effects that climate change is having on their economy, industries, and agriculture. These include increasing droughts in northern regions, which are delaying their efforts to eradicate poverty and to meet the Millennium Development Goals. To date, most if not all of these mitigation policies have been driven primarily by non-climate related objectives such as economic growth, improved air quality and energy security (Morse *et al.*, 2010). There is also an urgency to address environmental pollution as industrial effluents have resulted in the deterioration of air and water quality in all segments of Chinese society, resulting in health problems and well-documented "cancer villages."[1] Indeed, these factors have helped China to lower their energy and emissions intensities. Tying co-benefits to mitigation measures is critical in designing programs that incentivize acceleration toward meeting global targets for carbon abatement.

In March 2011, China released the draft of its 12th FYP, which for the first time explicitly mentions climate change and its commitment to the international climate negotiation process (Xinhua, 2011). The draft also calls for a 16 percent reduction in energy intensity and 17 percent reduction in CO_2 emissions per GDP by 2015, which is in line with COP15 and COP16 commitments to reduce the intensity of carbon dioxide emissions per unit of GDP in 2020 by 40 to 45 percent from 2005 levels.

Energy production

China currently relies on coal to meet two-thirds of its energy needs, with 80 percent of electricity being generated from this fossil fuel (Asia Society, 2009). In spite of government policies to promote hydropower, nuclear power, and renewable energy, and the recognition of the highly polluting nature of coal, its share in electricity production has grown in recent years. In 2006 and 2007 alone, China added about 170 GW of coal-based capacity, more than in the previous six years. Today there are more coal-fired power plants in China than in the United States, the United Kingdom, and India combined. Moreover, China's coal power use is expected to more than double by 2030, generating an additional 86 billion t-C (EIA, 2007).

In order to simultaneously address its growing economy, energy security, and poverty and contribute to the future climate regime, China is taking a multipronged approach of developing and deploying renewable energy, nuclear, clean coal technologies, transportation technologies, and energy efficiency across the industrial and buildings sectors of the economy.

Renewables

Hydropower generated from large dams and micro-turbines is the largest source of renewable energy in China. In 2009, China's hydroelectric generation capacity reached 172 GW, about one-quarter of its full potential. Under the 11th FYP, hydropower is expected to reach 300 GW by 2020.

In order to accelerate the rate of renewable energy penetration in China, the National Renewable Energy Law adopted a renewable energy target of 20 percent for the nation's electricity and 16 percent for its primary energy. To spur development in these new energy sources, incentives have included a national fund and discounted lending and tax preferences for renewable energy projects. The Chinese government has identified these "clean technologies," which include solar, wind, and energy storage, as key to their economic competitiveness, and has implemented policies for domestic innovation and industries, for example, by requiring the use of domestically manufactured parts. In 2009, China's investment in clean technologies exceeded that of the United States and took the top position in financing with $54.4 billion (Pew, 2011).

As a result of China's aggressive push in building their renewable energy industries, it is now the largest wind energy provider in the world with domestic capacity at 43 GW at the end of 2010. Through feed-in tariffs (FITs), the Chinese government has been able to double domestic installed capacity each year between 2005 and 2009 (Figure 6.2). As of 2011, China accounts for 20 percent of global wind capacity.

China is also the world's largest producer of solar photovoltaic cells, accounting for 35 percent of the global market. During the 11th FYP, the government funded solar demonstration projects for both heat and electricity production to drive market adoption. Yet, domestic installations of photovoltaic panels have been small due to their high costs. Today, solar power only provides 0.01 percent of its electricity. On the other hand, China has seen better success with solar thermal heaters and has manufactured and installed more solar water heaters than the rest of the world combined.

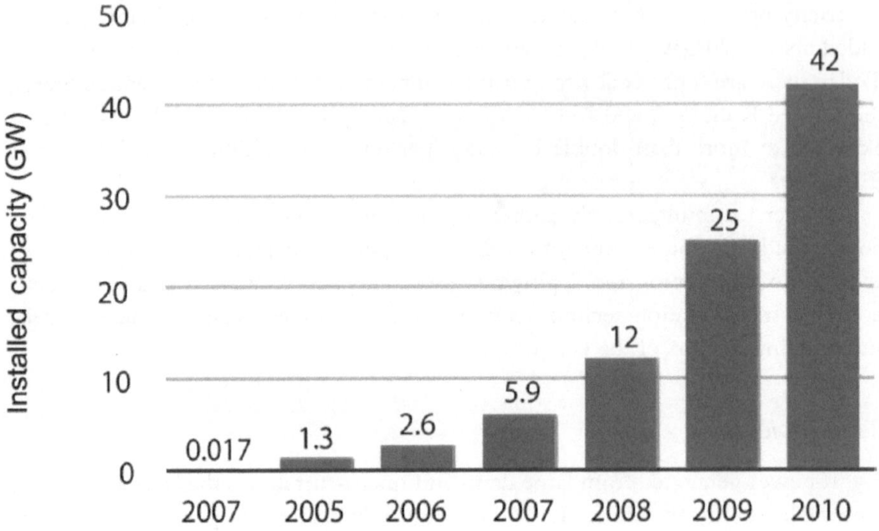

FIGURE 6.2 Growth of China's installed wind capacity (source: adapted from EIA International Energy Statistics, 2011).

Coal

In spite of the potential for renewable energy to provide carbon-free electricity and for energy efficiency to reduce energy demand, coal will be needed to power China's economy for decades to come. In the buildings and industrial sectors, most electricity consumed in Asia is produced from these plants, which are the largest sources of regional and global emissions (MIT, 2007). Coal is not only the least expensive option, it is more easily scalable than renewable energy. The development of clean coal technologies like carbon capture and storage (CCS) can potentially be an inexpensive approach to lower CO_2 emissions at a cost of $10 per t-$CO_2$. The IPCC notes that emissions from coal could be reduced by up to 90 percent using CCS if the technology can be proven and scaled up. Although widespread deployment of CCS is not to be expected anytime soon, preparations to facilitate their proper development by anticipating future infrastructure requirements will greatly enhance its introduction (Liu and Gallagher, 2009).

In addition to developing advancing low emission coal technologies, China will also need to commit to regulating emissions from coal-fired plants. While relatively stringent regulations to reduce coal related emissions from existing plants have been mandated, reductions of emissions from existing plants have met with limited success as enforcement is often lax and compliance low. This is because the operation of pollution control equipment reduces the net output of power plants, decreasing revenue for power producers. Instead, shutting down inefficient plants has been more effective, and, in the last several years regional authorities have stepped in to shut down more than 1,000 older coal-fired power plants and inefficient factories.

Industry

As the world's factory, China's emissions are dominated by heavy industries. Today, about 35 percent of the world's steel, 50 percent of its cement, and 28 percent of manufactured aluminum are produced in China. Moreover, steel alone emits more CO_2 than all Chinese households combined, the chemical industry uses more energy than all cars on the road, and aluminum smelters consume more electricity than the entire commercial sector (Houser *et al.*, 2008). While some of these outputs are exported, the vast majority is consumed domestically.

The Ten Key Energy Conservation Projects were established to promote energy efficiency among industrial enterprises through optimization of the economic structure, supporting advanced technologies, establishing strict management systems, and implementing effective incentive mechanisms (Price *et al.*, 2011). Specifically, the Ten Key Projects focused on coal-fired industrial boiler (kiln) retrofits, district cogeneration, waste heat and pressure utilization, petroleum conservation and substitution, motor energy efficiency, and energy system optimization (NDRC, 2006).

Another initiative for realizing China's 20 percent energy intensity reduction goal was the Top-1000 Energy-Consuming Enterprises program, which set energy conserving targets for 1,000 of the highest energy-consuming enterprises. Launched in 2006, the Top-1000 program encompassed large-scale enterprises in nine energy intensive industries with the goal of reducing unit energy consumption for all major products to either international standards or national leading levels (NDRC, 2006). It also called to achieve energy savings of 2.93 EJ in the 11th FYP period.

The Top-1000 enterprises are expected to establish an energy conservation organization, formulate energy efficiency goals, establish an energy utilization reporting system, conduct energy audits, conduct training, formulate an energy conservation plan, adopt energy conservation incentives, and invest in energy efficiency improvement options. The enterprises are required to report their fuel consumption quarterly to the National Bureau of Statistics (NDRC, 2006).

Due to rapid implementation, program targets were established without detailed assessments. The program goal represents only 15 percent of total required energy savings in the 11th FYP, yet the Top-1000 enterprises represent the highest energy consumption in the economy. A more ambitious goal likely could have been set based on assessment of potential savings in industrial subsectors.

Buildings

As hundreds of millions of Chinese move to cities and adopt modern lifestyles, they will undoubtedly use more electronics and appliances. By 2005, China's 190 million urban households owned an average of 1.3 color televisions, a clothes washer, refrigerator, and air conditioner (NBS, 2009). In addition, the ownership rate of personal computers grew from zero to over 40 percent in seven years (Fridley *et al.*, 2007). The rapid adoption of electric appliances has been identified as one of the drivers of residential electricity consumption, which grew at 13.9 percent annually between 1980 and 2007.

To ensure that these devices use energy efficiently and help to slow China's growing appetite for electricity, minimum energy performance standards (MEPS), which mandate maximum allowable energy consumption for 30 types of appliances and equipment, have been instituted. Subsequent MEPS revisions typically increased stringency by about 10 percent over the previous level. At the same time, the Top Key Energy Conservation project has also contributed to reduced emissions in the buildings sector through building energy conservation, green lighting, and government agency energy conservation (NDRC, 2006).

Transportation

Among developing countries, China has been a leader in adopting fuel economy standards (FES) for personal vehicles (Oliver *et al.*, 2009). Although these efforts are driven by energy security concerns, there is also a genuine desire among the Chinese to see their own auto companies not only become competitive against foreign companies in the domestic market but eventually in the international market. In 2004, China issued its FES for light-duty passenger vehicles (LDPV). In 2008, these standards were revised and now rank third globally, following the Japanese and European standards. This has resulted in a reduction of the average fuel consumption of the new national LDPV fleet by 11.5 percent.

India

India's economy experienced dramatic growth in past decades, lifting hundreds of millions out of poverty. Between 1981 and 2005, the poverty rate dropped from 60 percent to 42 percent. This transformation has been accompanied by a commensurate increase in energy demand to meet its growing industries and urban populations. Much like the Chinese economy, India's energy supplies have been met largely through fossil fuels, namely coal, which is its largest source of emissions.

With climate change already having severe impacts including coastal erosion, seawater intrusion, floods, droughts, crop failures, and disease outbreaks (Cruz *et al.*, 2007), leaders in India recognize the urgency to create both mitigation and adaptation strategies in their agricultural, health, and forestry sectors but without compromising their priorities to further reduce poverty. India has made substantial progress in reducing its poverty from 51 percent in 1990 to a target of 22 percent by 2015 (UN, 2010). The lack of access to efficient energy services is a major barrier to achieving vital and long-lasting solutions to poverty. Like China, India has continued to resist international pressures to agree to limit their emissions, due to their lack of "historical responsibility," which only account for 2 percent of accumulated global emissions.

As the world's fourth largest economy and fourth largest GHG emitter as of 2011 (UN, 2011), India currently accounts for about 4 percent of global emissions. Although it is a rapidly growing economy, India is still one of the least carbon intensive countries in the world. Although its emissions increased 65 percent between 1990 and 2005, India's per capita emissions are 70 percent below the world average and 93 percent below those of the United States. Nevertheless, current rates of economic growth are driving India to dramatically increase its demand for energy and consequently its emissions, which are expected to grow by 70 percent from 2005 levels by 2020.

India's participation in the international climate regime will not only have implications for global GHG emissions but also for climate resilience and security in the Southeast Asia region. While India is not obligated to reduce GHG emissions under the current United Nations Framework Convention on Climate Change (UNFCCC), it has begun considerable efforts to reduce GHGs without sacrificing its right to development. Indeed, it has been a leading voice of the developing world's perspective on climate change following COP8 in New Delhi. In its submission to the UNFCCC in January 2010, India committed to a reduction in emissions intensity of 20 percent to 25 percent from 2005 levels by 2020 with the condition that international support will be available (Government of India, 2010).

In July 2007, the Prime Minister's Council on Climate Change (PMCCC) was established to coordinate with various stakeholders in developing a broad national-level response to climate change and to provide oversight on key policy decisions, including the formulation of climate assessment, mitigation, and adaptation plans. In July 2008, Prime Minister Manmohan Singh released India's first National Action Plan on Climate Change (NAPCC), which contains eight national missions with the aim of mitigating and adapting to climate change, but in a manner that also addresses the parallel concerns of energy security and poverty (Government of India, 2008). These include:

- protecting the poor and vulnerable sections of society through an inclusive and sustainable development strategy, sensitive to climate change;
- achieving national growth objectives through a qualitative change in direction that enhances ecological sustainability, leading to further mitigation of greenhouse gas emissions;
- devising efficient and cost-effective strategies for end-use Demand Side Management;
- deploying appropriate technologies for both adaptation and mitigation of greenhouse gases emissions extensively as well as at an accelerated pace;
- engineering new and innovative forms of market, regulatory, and voluntary mechanisms to promote sustainable development.

These efforts will be in effect through 2017, directing the ministries to submit detailed implementation plans to the Prime Minister's Council on Climate Change.

In 2010, the Planning Commission created an Experts Group to prepare a strategy for a low carbon economy that could feed into its 12th FYP process, which will start in 2012.

Although the goals outlined in the national missions are commendable, there are very few details as to how they would be achieved. There has been some criticism of the piecemeal approach of the missions and the lack of a comprehensive and coherent set of strategies to deal with the multidimensional nature of the climate challenge. In addition, these initiatives are often personality driven and at the institutional level there is little capacity for analysis and informed policy making.

Nevertheless, domestic reforms to improve energy efficiency, increase renewable and nuclear energy, expand public transportation and restructure energy price over the past few decades have resulted not only in higher electrification rates, but also resulted directly in the decline of per capita carbon intensity of the Indian economy (Government of India, 2008). While these efforts were not originally designed with sustainable development in mind, they are contributing to global efforts to mitigate GHG emissions.

Energy production

Coal use contributes to about two-thirds of India's CO_2 emissions, mainly from the power, steel, and cement sectors, while biomass burning and agriculture are the main sources of methane and N_2O emissions (Garg *et al.*, 2006). For power generation, coal-based generation capacity contributes to about 55 percent, followed by hydro and natural gas. The dominance of coal, which is the largest source of CO_2 emissions, is expected to continue into the future (TERI, 2006). Driven by economic and population growth, India will need to import additional coal supplies to meet growing demand for energy. India currently imports about 10 percent of its coal for electricity generation, which is projected to reach 16 percent by 2011 (Sargsyan *et al.*, 2010). Indeed, India's growing reliance on coal is not only unsustainable, as its coal reserves are expected to run out in 45 years, it is also a threat to its energy security.

Renewable energy

India has long recognized the need to develop non-hydro renewable sources of energy, which now account for 3.5 percent of energy generation, as a way to diversify energy supplies, reduce dependence on foreign energy sources, and mitigate fuel price volatility. In 1992, the Ministry of Non-Conventional Energy Sources (renamed in 2006 as the Ministry of New and Renewable Energy) launched research and development (R&D) programs and helped engineer a shift from subsidy-driven dissemination initiatives to the commercialization of low carbon technologies. Furthermore, the Electricity Act 2003 and the National Tariff Policy 2006 mandated central and state electricity regulatory commissions to purchase a certain percentage of grid-based power from renewable sources (Pew, 2008).

Furthermore, the Ministry has also set a goal of using renewables for 10 percent of new power generation capacity by 2010. India's policies for developing wind and solar power include preferential tariffs as well as fiscal incentives such as accelerated depreciation and renewable portfolio standards (RPS).

Current estimates now suggest 150 GW of renewable energy potential, about half in the form of small hydropower, biomass, and wind, and another half from solar, co-generation, and waste-to-energy that can be developed in India (Sargsyan *et al.*, 2010). Of that total, only 10 percent or 15 GW has been developed and the national government has now set a target of increasing renewable energy output by 40 GW to coincide with the target of 55 GW at the end of the 13th FYP in 2022.

Under its National Solar Mission (Government of India, 2008), India aims to develop and use solar energy for power generation and other uses with the ultimate objective of making solar competitive with fossil-based energy options. The plan includes increasing production of photovoltaics to 1,000 MW/year by 2013 and deploying at least 1,000 MW of solar thermal power generation in urban areas, industry, and commercial establishments. Contingent upon the success of this initial phase, the government plans to increase annual production to 20,000 MW by 2020. Additionally, the mission calls for the establishment of a solar research center, increased international collaboration on technology development, strengthening of domestic manufacturing capacity, and incentives for domestic and foreign investment (Pew, 2008).

The government now supports solar thermal projects through capital subsidies, sales incentives, and reimbursement of fees. For solar photovoltaic technologies, foreign investors are offered automatic approval procedures for up to 74 percent of foreign direct investment in joint venture projects and up to 100 percent if approved by the Foreign Investment Promotion Board.

While solar is the only renewable energy source that is explicitly spelled out as a National Mission, entrepreneurship from the private sector and various government policies have been supporting other low carbon technologies. Although India only started to build its wind market during the 1990s, it had the world's fifth largest installed capacity as of 2010, reaching 13 GW, more than two-thirds of all renewable power generation. Due to the incentives offered for wind power, India is the world's third largest wind energy installer, capturing nearly 10 percent of the total global market, and is home to Suzlon, a major company in the global wind market. This dramatic growth is partly attributed to a policy framework that reduced upfront costs to investors (Pew, 2008). The government's Renewable Energy Development Agency has been providing long-term low-interest loans, and investors are allowed 80 percent accelerated depreciation for the first year; concessions on import duties, sales tax, and excise duties; and a ten-year income tax exemption for profits from wind generation. Moreover, subsidies have also been provided for demonstration purposes. Nevertheless, many operating plants are generating far less than their full capacity because current incentives encourage the installation of wind generators but not their operation.

Furthermore, India has one of the largest resources for hydropower, which is already contributing to about a quarter of India's energy needs (ADB, 2007b). There is a potential for 150 GW of conventional hydropower generation of which only 32 GW has been developed. In spite of the enormous potential to further provide clean energy, economic and political barriers have limited the development of large-scale hydroelectric plants.

For biomass energy production and hydropower energy development, various incentives have also been available in the form of subsidies, tax holidays, excise duty and sales tax exemptions, and accelerated depreciation (Pew, 2008). Although relatively "low-tech" compared to solar and wind technologies, hydro and biomass are among the most affordable options with an average economic cost of 3.6 IND/kWh and 4.6 IND/kWh, respectively. Moreover, biomass technology is appropriate in rural economies that are not connected to the grid and have access to biomass feedstock. For hydropower, there is still enormous potential for small systems (less than 25 MW) to be developed, particularly in rural areas, where less intensive, low-tech solutions may be the ideal form of energy development.

Undoubtedly, the cost of deploying renewable energy at this scale will be both a financial and technical challenge. An analysis from the World Bank suggests the key to a successful shift to renewable energy lies in the realignment of subsidies away from fossil fuels to renewable sources; in other words, a phasing out of the wrong incentives. Furthermore, in a report from the Energy Sector Management Assistance Program (ESMAP) of the World Bank, the cost of reaching the government target will be between $10 and $64 billion (Sargsyan *et al.*, 2010). The lower-cost figure assumes the deployment of low-diversity, low-cost renewable energy sources, and the higher-cost figure is based on a diverse mix of renewable sources, including more expensive solar power. A simple analysis shows that the low-cost option is attainable when costs are spread over the period of implementation. Additionally, the higher-cost scenario becomes attractive if fossil fuel prices continue to rise in the absence of distorting subsidies.

For renewable energy to truly fulfill its potential in providing clean power, an advanced national grid needs to be developed so that power can be efficiently transmitted and distributed from energy surplus regions to underserved and remote areas that currently have unreliable or no access to electricity. This infrastructure will enable sources like small hydropower, which are abundant yet untapped, to supply clean and stable power across the country. In 2009, the World Bank provided a $1 billion loan to India's national power utility to strengthen five transmission systems in the northern, western, and southern regions and to improve overall transmission of electricity.

Coal power

Like China, India will continue to rely on coal power for many years. Among the low hanging fruits for reducing GHG emissions is retirement of inefficient coal-fired power plants, which account for 7 percent of installed coal capacity.

According to the National Mission for Enhanced Energy Efficiency, the government is mandating their retirement by 2012 with an additional 10 GW that will be retired or reconditioned by 2017. India has also been supporting the research and development of advanced combustion technologies since 2002 under its 10th FYP (Government of India, 2008), and has plans to build three Integrated Gasification Combined Cycle (IGCC) plants for demonstration and is actively encouraging the use of supercritical technologies.

Industry

India's industries account for 42 percent of commercial energy. In 2001, the Energy Conservation Act established a national Bureau of Energy Efficiency (BEE) to improve energy efficiency in various sectors, which include large energy consuming industries (Government of India, 2008). Among its activities, BEE has conducted energy audits and established task groups for the textile, cement, pulp and paper, fertilizer, chlor-alkali, and aluminum sectors (Sathaye *et al.*, 2006). Participants from industry come to share their best practices, declare their voluntary targets, and adopt benchmarks. For example, a benchmarking tool is being developed by the Indo-German Energy Efficiency and Environment Project to enable cement manufacturers to determine their energy consumption levels relative to peers in the industry. These activities help to build capacity, encourage and mandate benchmarking at the plant level, and, in the long term, to establish standards and procedures for energy management in Indian industries.

The National Mission for Enhanced Energy Efficiency has built upon these existing programs to require specific energy consumption decreases in large energy-consuming industries, with a system for companies to trade energy-savings certificates. It also calls for the financing of public–private partnerships to reduce energy consumption through demand-side management programs in the municipal, buildings, and agricultural sectors.

Buildings

Although India's commercial and residential buildings sectors combined consume less power than its industrial sector, electricity use in buildings is skyrocketing. By 2008, total electricity usage in buildings increased to 33 percent from 24 percent in 1994 (Sathaye *et al.*, 2009). Among the earliest measures to improve energy-saving in buildings include the National Tariff Policy (2006), which implemented a higher tariff base for consumers in excess of 1 MW, and the Energy Conservation Building Code (2007), which established energy performance requirements for commercial buildings with loads of 500 kW and above.

The barriers hindering the market penetration of energy efficient appliances in the buildings sector include the lack of informed use of energy, lack of information to compare products, the lack of availability of more efficient products, and higher costs. There is also a perceived risk due to the lack of confidence in new technologies in appliances and buildings. To surmount these challenges, BEE, which also works with the industrial sector, has also developed demand-side management programs (DSM) to help overcome perceived risk, and standards and labeling to bolster energy efficient refrigerators and appliances to enable consumers to make informed choices. BEE has also instituted minimum energy performance standards for notified equipment and appliances.

Under the NAPCC, the National Mission on Sustainable Habitat calls for further improvements in the energy efficiency of buildings in the residential and commercial sectors (Government of India, 2008). The NAPCC estimates that implementing energy efficient options could help achieve 30 percent electricity savings in new residential buildings and 40 percent in new commercial buildings. Retrofitting existing buildings offers corresponding savings of 20 percent and 30 percent, respectively.

More recently, BEE has been involved in developing the Bachat Lamp Yojana program to lower the cost of energy efficient compact fluorescent lamps to the same level as conventional incandescent lightbulbs, which consume far more power (Pew, 2008). As a CDM project, the government expects to save 24 million t-CO_2 each year and to recover the price difference through carbon credits. It is projected that this program will cover all of India by 2012.

Transportation

Along with buildings, demand for personal vehicular transportation is growing dramatically in India, leading to increased air pollution, traffic congestion, and respiratory illnesses. Although these concerns are not directly linked to climate change, co-benefits generated by pursuing climate goals are needed. To manage the exploding demand for transportation, the Indian government has been setting vehicle standards and encouraging alternative fuels and forms of transportation (Pew, 2008). In 2003, the National Auto Fuel Policy was enacted to require all new four-wheeled vehicles in 11 cities to meet Bharat Stage III emission norms (similar to Euro III standards) for conventional air pollutants and to comply with Euro IV standards by 2010.

In New Delhi and Mumbai, the largest commercial fleet of compressed natural gas (CNG) vehicles was introduced to reduce pollution and to increase energy security. Of these vehicles, 106,000 buses, taxis, and three-wheelers were converted from gasoline or diesel to CNG in New Delhi. As other cities participated, a total conversion of 375,000 vehicles was achieved by March 2007 with three-wheelers taking the largest share (64 percent).

Despite the potential of various technologies to improve mileage and spur a switch away from conventional fossil fuels for transportation, the development of public modes of transportation will also be needed to mitigate GHG emissions and also to contribute to energy security, reduction of pollution, and various health benefits. One of the major public projects under construction is the Delhi Metro subway system, which began in 1998 and is expected to cover the entire metropolis by 2021. Although only the first phase has been completed, the daily ridership reached 1.6 million in 2010. The project is partly funded by the Japan International Cooperation Agency (JICA) and demonstrates the role of Overseas Development Assistance (ODA) in creating a mutually beneficial partnership and in realizing a co-benefits approach in facilitating environmentally sustainable transport.

Other sectors

In India, the agricultural and waste management sectors are also major sources of GHG emissions, namely in the form of methane gas. Although technologies to capture and use methane are readily available, they have not yet been brought to commercial scale. Composting has been found to be highly adaptable and suitable for community-driven initiatives and is readily scalable for cities in the Asia-Pacific region. Not only are GHG emissions reduced, waste is minimized at the same time. The extraction of energy and raw materials from waste also meets the sustainability goals of "reduce, reuse, recycle" (3Rs).

Way forward

Both China and India will need to alter their emissions trajectory if the world is to meet its target of not increasing the average global temperature by 2 degrees Celsius. Although neither has agreed to absolute caps on their emissions, they have submitted targets to reduce their energy intensities with the understanding that technologies and financing will be available from the developed world. Domestically, policies in both countries have led to periods of decarbonization in their economies.

Nevertheless, they continue to experience rapid growth and improving standards of living while being challenged by multiple priorities for energy security, climate vulnerability, and poverty. China and India along with other developing economies in Asia will need to properly address their lack of capacity to formulate effective climate responses and to overcome the notion that mitigation and adaptation efforts are not compatible with sustainable development. In enhancing the national and international actions for mitigation, the policy discussions have focused on overcoming inconsistent standards and protocols that are hindering effective financing, technology transfer, and capacity building from developed nations.

Technologies

Technologies provide options to enable countries to become less energy intensive and to make their energy supplies less carbon intensive. In India, renewable energy, biomass, cogeneration, and carbon capture have significant abatement potential (Ghosh *et al.*, 2002). In China, offshore wind, solar energy, nuclear power generation, fuel cells, hybrid vehicle, integrated gasification combined cycle (IGCC), clean coal technologies, and the advanced gas turbine are believed to be critical technologies to achieve LCS (Jiang, 2005).

Over the next half-century, China and India will increase their level of energy consumption up to sixfold. With oil likely to be depleted over this time period, the main options for affordable electricity will come from coal or nuclear power. In order to properly meet its current and future energy needs and to address its growing concern over meeting climate goals, both countries have already initiated policies to spur the development of various low carbon energy technologies including advanced coal power, renewables, energy efficiency, smart electricity grids, and sustainable transportation systems. There is an urgent need for them to collaborate with industrialized countries and among themselves to accelerate the deployment of technologies.

- *Advanced coal technologies*. As the cheapest and easily scalable source of energy, coal power will supply the majority of Asia's power for decades to come. There is an enormous need to develop advanced coal technologies to improve efficiency and to also capture GHG emissions using CCS (MIT, 2007). Given the complexity and enormous costs of developing, demonstrating, and maintaining CCS technology, Asian countries, with China in the lead, should work with developed countries in assessing the cost, viability, and effectiveness of this potentially breakthrough technology (Asia Society, 2009).
- *Energy efficiency*. China, India, and other Asian countries should continue in scaling up existing collaborations with the US, exchanging know-how on best practices for energy efficiency standards in the residential and commercial sectors, as well as enabling energy intensive industries to benchmark their emissions. They should also maintain their ongoing work in incentivizing small and medium enterprises to abide by efficiency standards and in developing fuel economy standards for personal vehicles. The public sector can also play a major role in driving momentum for energy conservation. Government programs in China to implement energy conservation have shown early successes that should be shared.
- *Renewable energy*. China and India have already demonstrated tremendous innovative capacity in manufacturing and are the world's largest producers of wind and solar systems. Yet, renewable energy is still prohibitively expensive for many people in Asia and the financial capacity and human resources needed to scale up and maintain these resources are not widely available in these areas. Thus, governments from developed countries need to step in and implement R&D policies to help make these technologies more affordable and scalable.
- *Smart electricity grids*. Because much of the infrastructure (buildings, power plants, and transportation systems) has yet to be built, there is an enormous opportunity for Asian economies to leapfrog over the carbon intensive patterns of development through the design and implementation of advanced electricity grids in the context of smart urban design. These systems will not only improve the overall efficiency in power transmission and distribution, they will enable renewable energy systems to realize their potential in contributing

to reliable, low emissions electricity. However, demonstration of fully integrated smart grid systems is still needed, and planned low carbon cities are ideal test beds for their experimentation.

- *Transportation.* In the transportation sector, the rapid growth of motorized transport is a leading cause of air pollution and GHG emissions. WRI (2004) recommends developing a framework for deploying low carbon transport across Asia as it rapidly urbanizes and becomes dependent on oil for transportation. There is also a need to integrate vehicle and engine design, fuels, and urban planning in order to properly develop transportation in a sustainable manner. A co-benefits approach – for example the New Delhi Metro project – can help to create opportunities for mutual benefits through international technical and financial assistance. Chinese efforts to spur vehicle innovation through mileage standards provide valuable experience for developing countries experiencing or anticipating rapid growth in personal vehicles.

Financial

The costs of transitioning to a low carbon future will be staggering. The IEA projects that between 2006 and 2030, China and India will need to invest \$2.8 trillion and nearly \$1.0 trillion in the power sector, respectively (IEA, 2007b). Also, at the global level, the UNFCCC (2007) estimates that an additional annual investment of \$200 billion is in order for GHG emissions to return to 2006 levels by 2030.

Both public and private funding will be critical in mobilizing low carbon development in Asia. Multilateral organizations have already contributed to financing low carbon energy infrastructure in Asian countries; however, they must be scaled up many times in order to make a meaningful impact. Examples of these instruments include the World Bank Climate Investment Fund, the newly proposed Clean Technology Fund (CTF), and the Global Environment Facility Trust Fund (Asia Society, 2009). Nevertheless, the critical challenge will be in using these and other public modes of financing to best leverage much larger sums from the private sector. The removal of tariffs and non-tariff barriers is still needed to promote the transfer of low carbon technologies for commercial development in Asia.

Under international climate negotiations, measurable, reportable, and verifiable (MRV) has emerged as a strategy to help developing countries make commitments to nationally appropriate mitigation actions (NAMAs) and link them to the enabling conditions (IGES, 2010). By pledging to a NAMA, a UNFCCC sanctioned organization uses the MRV criteria to quantitatively determine the appropriate levels of technology, financing, and capacity building that developing countries should send to the applicant. The verified emission reductions achieved by NAMAs would then give developed countries credits for compliance with their commitments (see Chapter 7).

In addition to developing the financial capacity needed to decarbonize their economies, China and India and other Asian economies must also contend with multiple climate challenges including floods, droughts, and other natural disasters with broad impacts on livelihood. Although funding for adaptation measures is gaining interest, there is an opportunity to bring down costs by identifying synergies between mitigation and adaptation measures.

Capacity building

In order to enable the appropriate uptake of technologies and to mobilize finances in the context of an economy's values, aspirations, and resource endowment, capacity must be built at the institutional, civic, and individual levels. These are facilitated through exchanges at the global, regional, and urban levels of governance.

International cooperation

While China and India were originally exempted from capping their emissions under the Kyoto Protocol due to their lack of historical responsibility, these two countries present both a global challenge and an opportunity to manage the climate problem. Neither country can tackle the energy and climate problem alone; collaboration will be required with the rest of the world in order to make the transition to low carbon economies. The world is at a critical juncture where both developing economies in Asia and the developed world are undergoing profound structural changes. In the post-2012 climate regime, legal, institutional, and financial frameworks must be developed while clarifying "common but differentiated responsibilities" between the developed and developing worlds and paying due attention to millennium development goals including equity, cost-effectiveness, environmental outcome, and flexibility in the future climate regime.

Regional

In consultations between policy makers and researchers among Asian economies and with developed countries, new perspectives have emerged to shape the future climate regime. The current policy dialogues are aimed at "(a) promoting new and constructive thinking in the Asia-Pacific region on future climate actions and (b) contributing to the shaping of a post-2012 climate regime that adequately reflects the concerns and developmental aspirations of countries in the Asia-Pacific region" (IGES, 2008).

Urban-level exchanges

In contrast to the national targets to reduce emissions, a bottom-up approach from the urban level of governance will also play a critical role in reducing GHG emissions. In both industrialized and developing economies, cities are increasingly becoming the centers of decision making. Urban initiatives can therefore take into account the wider resource, socioeconomic, and geopolitical diversity that exists among cities. Indeed, with decentralization being a trend in developing countries, efforts to promote smart buildings, environmentally sustainable transportation, and energy efficiency at the urban level may play a greater role than national directives to mitigate emissions.

City-to-city exchanges through organizations like C40, ICLEI, and CITYNET enable urban leaders to learn about and adopt best practices for GHG mitigation. These partnerships go beyond individual city-to-city collaborations, becoming a network of small or medium sized urban areas (see Chapter 5). Non-governmental organizations (NGOs) serve a critical role at the local level and can help to build linkages and exchanges between different stakeholders in government, industry, academia, and civic society. At the same time, universities and institutions in Asia can be a catalyst in improving access to education and research opportunities, especially for students and young researchers. Developing a common standard for communication will improve the exchange of information and ideas.

Acknowledgments

The author wishes to thank Professor Jusen Asuka for kind support and useful comments for this chapter.

Note

1 Cancer villages have been widely reported in the media as villages in China where the number of cancer patients is abnormally high. See article at www.environmentmagazine.org/Archives/Back%20Issues/March-April%202010/made-in-china-full.html.

References

Adger, W.N., Hughes, T.P., Folke, C., Carpenter, S.R., Rockström, J. (2005) "Social-Ecological Resilience to Coastal Disasters," *Science*, 309 (5737), 1036–1039.

Asian Development Bank (ADB) (2007a) "News Brief: Conference to Explore Cooperation on Asia's Urban Challenges," www.adb.org/media/Articles/2007/11427-asian-urbanizations-conferences/ (accessed May 10, 2009).

Asian Development Bank (ADB) (2007b) *Hydropower Development in India: A Sector Assessment*, Manila, Philippines: ADB.

Asia Society (2009) "Common Challenge, Collaborative Response: A Roadmap for U.S.–China Cooperation on Energy and Climate Change," Asia Society Center on U.S.–China Relations, Pew Center on Global Climate Change.

Cruz, R.V., Harasawa, H., Lal, M., Wu, S., Anokhin, Y., Punsalmaa, B., Honda, Y., Jafari, M., Li, C., and HuuNinh, N. (2007) "Asia," *Climate Change 2007: Impacts, Adaptation and Vulnerability, Contribution of Working Group II to the Fourth Assessment Report of the Intergovernmental Panel on Climate Change* [M.L. Parry, O.F. Canziani, J.P. Palutikof, P.J. van der Linden, and C.E. Hanson, Eds.], Cambridge: Cambridge University Press, 469–506.

EIA (Energy Information Agency) (2007) Chapter 5, *International Energy Outlook*, US Department of Energy, www.eia.doe.gov/oiaf/ieo/index.html (accessed May 1, 2007).

Fridley, D., Aden, N., Zhou, N., and Lin, J. (2007) *Impacts of China's Current Appliance Standards and Labeling Program to 2020*, Berkeley, CA: Lawrence Berkeley National Laboratory.

Fujino, J., Hibino, G., Ehara, T., Matsuoka, M., Masui, T., and Kainuma M. (2008) "Backcasting Analysis for 70% Emission Reduction in Japan by 2050," *Climate Policy*, 8 (1), S108–S124.

Garg, A., Shukla, P.R., and Kapse, M. (2006) "The Sectoral Trends of Multigas Emissions Inventory of India," *Atmospheric Environment*, 40 (24), 4608–4620.

Ghosh, D., Shukla, P.R., Garg, A., and Ramana, P.V. (2002) "Renewable Energy Technologies for the Indian Power Sector: Mitigation Potential and Operational Strategies," *Renewable and Sustainable Energy Reviews*, 6 (6), 481–512.

Government of India (2008) *National Action Plan on Climate Change*, Prime Minister's Council on Climate Change.

Government of India (2010) Letter to the UNFCCC, Bonn, Germany, January 30, http://unfccc.int/files/meetings/cop_15/copenhagen_accord/application/.

Hanaoka, T., Akashi, O., Kanamori, Y., Hasegawa, T., Hibino, G., Fujiwara, K., Kainuma, M., and Matsuoka, Y. (2008) "Global Greenhouse Gas Emissions Reduction Potentials and Mitigation Costs in 2020: Methodology and Results." *CGER-Report*, Tsukuba, Japan: Center for Global Environmental Research, National Institute for Environmental Studies.

Houser, T., Bradley, R., Childes, B., Werksman, J., and Keilmayr, R. (2008) *Leveling the Carbon Playing Field: International Competition and US Climate Policy Design*, Washington, DC: Peterson Institute for International Economics and World Resources Institute.

IEA (International Energy Agency) (2007a) *Tracking Industrial Energy Efficiency and CO_2 Emissions: In support of the G8 Plan of Action*, Paris: IEA.

IEA (International Energy Agency) (2007b) *World Energy Outlook*, Paris: IEA.

IEA (International Energy Agency) (2011) *World Energy Outlook 2011*, Paris: IEA.

IGES (2008) "Consultations on the Post-2012 Climate Regime: Round 4," www.iges.or.jp/en/news/press/08_08_13.html (accessed April 30, 2009).

IGES (2010) *Measurable, Reportable and Verifiable (MRV): Verifiable (MRV) Trends and Developments in Climate Change Negotiations*, Hayama, Japan: IGES.

IPCC (2001) *Climate Change 2001: Mitigation* [B. Metz *et al.* (Eds.)], Geneva: IPCC Secretariat, www.grida.no/climate/ipcc_tar/wg3/index.htm (accessed March 23, 2009).

Jiang, K. (2005) "Technology Development for China: Energy and Climate Change" presented at the GTSP meeting, Pacific Northwest National Laboratory (PNNL), Richland, WA, May 24.

Lewis, J. (2007) "China's Strategic Priorities in International Climate Change Negotiations," *Washington Quarterly*, 31 (1), 155–174.

Liu, H. and Gallagher, K.S. (2009) "Catalyzing Strategic Transformation to a Low-Carbon Economy: A CCS Roadmap for China," *Energy Policy*, 38 (1), 59–74.

MIT (Massachusetts Institute of Technology) (2007) *The Future of Coal*, Cambridge, MA: An Interdisciplinary MIT Study.

Morse, R.K., He, G., and Rai, V. (2010) "Carbon Capture and Storage Technology in China is Driven by Energy Security Concerns" *BusinessForum China*, February, 30–32.

NBS (National Bureau of Statistics) (2009) *China Statistical Yearbook*, Beijing: NBS.

NRDC (National Development and Reform Commission) (2006) "Implementation Suggestions of Ten Key Energy-Conservation Projects during the Eleventh Five-Year Plan" NDRC Department of Resource Conservation and Environmental Protection Document #: [2006]1457.

Oliver, H.H., Gallagher, K.S., Tian, D., and Zhang, J. (2009) "China's Fuel Economy

Standards for Passenger Vehicles: Rationale, Policy Process, and Impacts," Discussion Paper 2009–03, Energy Technology Innovation Policy Research Group, Belfer Center for Science and International Affairs, Harvard Kennedy School, Cambridge, MA.

Pacala, S. and Socolow, R. (2004) "Stabilization Wedges: Solving the Climate Problem for the Next 50 Years with Current Technologies," *Science*, 305 (5686), 968–972

Pew (2008) "Climate Change Mitigation Measures in India," *International Brief*, Virginia, USA: Pew Center of Global Climate Change.

Pew (2011) *Who's Winning the Clean Energy Race? Growth, Competition and Opportunity in the World's Largest Economies*, California, USA: Pew Charitable Trusts.

Price, L., Levine, M.D., Zhou, N., Fridley, D., Aden, N., Lu, H., McNeil, M., Zheng, N., Qin, Y., and Yowargana, P. (2011) "Assessment of China's Energy-Saving and Emission-Reduction Accomplishments and Opportunities during the 11th Five Year Plan," *Energy Policy*, 39 (4), 2165–2178.

Sargsyan, G., Bhatia, M., Banerjee, S.G., Raghunathan, K., and Soni, R. (2010) *Unleashing the Potential of Renewable Energy in India*, Washington, DC: World Bank.

Sathaye, J.A., Bharvirkar, R., de la Rue du Can, S., Ghatikar, G., Iyer, M., Masanet, E., Price, L., Vine, E., Kumar, S., Kapoor, R., and Padmanabhan, S. (2009) *Energy Efficiency and Sustainable Development: Potential for US–India Collaboration in Buildings, Industry and the Smart Grid*, Berkeley, CA: Lawrence Berkeley National Laboratory.

Sathaye, J.A., de la Rue du Can, S., Kumar, S., Iyer, M., Galitsky, C., Phadke, A., McNeill, M., Price, L., Bharvirkar, R., and Padmanabhan, S. (2006) *Implementing End-Use Efficiency Improvements in India: Drawing from Experience in the US and Other Countries*, Berkeley, CA: Lawrence Berkeley National Laboratory.

TERI (The Energy and Resources Institute) (2006) "National Energy Map for India, Technology Vision 2030," PSA/2006/3, Study by TERI for Office of the Principal Adviser to the Government of India.

UN (United Nations) (2010) "Millennium Development Goals Report 2010," www.un.org/millenniumgoals/pdf/MDG%20Report%202010%20En%20r15%20-low%20res%2020100615%20-.pdf (accessed November 30, 2011).

UN (United Nations) (2011) "Millennium Development Goals Indicators: Carbon Dioxide Emissions (CO_2)," http://mdgs.un.org/unsd/mdg/SeriesDetail.aspx?srid=749&crid (accessed July 30, 2011)

UNFCCC (United Nations Framework Convention on Climate Change) (2007) *Investment and Financial Flows to Address Climate Change*, http://unfccc.int/resource/docs/publications/financial_flows.pdf (accessed August 16, 2011).

WRI (World Resources Institute) (2004) "Sustainable Urban Transport in Asia: Making the Vision a Reality," WRI, Washington, DC.

Xinhua (2011) "Key Targets of China's 12th Five-Year Plan" *Xinhua News Agency*, March 5, www.chinadaily.com.cn/xinhua/2011–03–05/content_1938144.html (accessed April 5, 2011).

7

DEVELOPMENT OF A NATIONALLY APPROPRIATE MITIGATION ACTION PLAN (NAMA)

A case study of Thailand's waste and wastewater management sector

Kazuhito Yamada and Mariko Fujimori

NAMAs and MRV in developing countries in international negotiations

In order to combat issues related to climate change, it is obvious that greenhouse gas (GHG) emissions need to be reduced not only in developed countries but also in developing countries. The issue of GHG mitigation actions by developing countries was the focus of intensive discussions at the COP13 in 2007, held in Bali, Indonesia. The concept of "nationally appropriate mitigation actions," called "NAMAs," proposed by several of the Annex I and non-Annex I countries was taken up at the conference. Ultimately, the terms of NAMAs and of measurement, reporting, and verification (MRV) in developing countries were officially stipulated for the first time in paragraph 1(b)(ii) of the Bali Action Plan at COP13, as follows (UNFCCC 2007):

- *Nationally appropriate mitigation actions* [NAMAs] by developing country Parties in the context of sustainable development, supported and enabled by technology, financing, and capacity-building, in a *measurable*, *reportable*, and *verifiable* [MRV][1] manner;

After three years of intensive international discussions, the NAMA concept was consolidated at COP16 in 2010, held in Cancun, Mexico. The main clause of the COP agreement was as follows (UNFCCC 2010):

- [COP] agrees that developing country Parties will take [NAMAs] in the context of sustainable development, supported and enabled by technology, financing, and capacity-building, aimed at achieving a deviation in emissions relative to "business as usual" emissions in 2020.

In the rationale clauses, they recognized that

> developing country Parties are already contributing and will continue to contribute to a global mitigation effort in accordance with the principles and provisions of the Convention, and could enhance their mitigation actions, depending on the provision of finance, technology and capacity-building support provided by developed country Parties

and reaffirmed that "social and economic development and poverty eradication are the first and overriding priorities of developing country Parties, and that the share of global emissions originating in developing countries will grow to meet their social and development needs."

The COP also decided to set up a registry to record NAMAs submitted by developing countries seeking international support and to facilitate matching of finance, technology, and capacity-building support to these actions. The registry will record and regularly update the information provided by Parties on:

a. NAMAs seeking international support;
b. support available from developed country Parties for these actions; and
c. support provided for NAMAs.

With regard to MRV, the COP decided that internationally supported mitigation actions[2] will be measured, reported, and verified domestically and will be subject to international measurement, reporting, and verification in accordance with guidelines to be developed under the Convention, and that domestically supported mitigation actions[3] will be measured, reported, and verified domestically in accordance with the guidelines.

As mentioned above, remarkable progress has been made with NAMAs and MRV in developing countries in international negotiations. However, at the present stage, the detailed design of NAMAs and MRV has only just been started by the COP, ad hoc working groups, and related bodies. Therefore, there is a limited common understanding among the Parties about modalities and procedures for implementing NAMAs and MRV in developing countries.

In this chapter, with the aim of accelerating NAMA planning development by developing countries, the authors develop and introduce a methodology for developing a NAMA plan, by showing an example of its application to an actual feasibility study, after a review of proposed NAMAs in general.

Overview of proposed NAMAs

Forty-four non-Annex 1 Parties had submitted NAMAs pursuant to the Copenhagen Accord[4] as of March 18, 2011. Table 7.1 shows typical examples of NAMAs. Each Party has submitted different kinds of targets. For example, some countries have set a unit target such as "per unit of GDP" (China). For the base year, some countries have set an exact year such as "below 2009 levels" (Marshall Islands); on the other hand, some countries have set "business as usual (BaU)" instead of an actual year (Korea, Mexico, Singapore, South Africa). Only a limited number of countries have declared numerical targets for GHG emission reduction including actual sectors in which emissions are to be reduced (e.g., Brazil and China). Other countries have raised their numerical targets without specifying either the actual basis for setting the target or the reduction method (e.g., India, Indonesia). Some countries have submitted only a qualitative explanation of activities without numerical targets (e.g., Afghanistan); on the other hand, some countries have submitted very detailed lists of projects to be implemented as their NAMA (e.g., Georgia, Ghana). Maldives and Bhutan have declared a goal of making themselves "100% carbon neutral" countries.

It is notable that most ASEAN countries, such as Cambodia, Thailand, and Vietnam, had not submitted their NAMAs as of March 2011. Among the ASEAN countries, only Indonesia submitted its NAMA before the first deadline (January 31, 2010). The Designated National Authorities (DNA) of Thailand have asserted that they will submit their NAMA by COP17.[5]

Key issues for developing a NAMA implementable with MRV

The four key issues outlined below should be addressed in order to develop a NAMA implementable with MRV.

What should the NAMA contribute to?

As mentioned above, the rationale clause reaffirmed that "social and economic development and poverty eradication are the first and overriding priorities of developing country Parties." Therefore, the modalities and procedures of NAMAs have to take sustainable social and economic development into account in implementing NAMAs. In order to implement appropriate NAMAs in developing countries, we first need to develop a NAMA planning methodology. The main objective in developing this methodology should not only be to contribute to GHG emission reduction but also to contribute to the sustainable development of the developing country Parties. The methodology should support the achievement of sustainable development in each sector, and simultaneously achieve reduction of GHG emissions related to the activities of these sectors. To this end, existing issues and problems in each sector should be identified as the first step. Countermeasures to address these issues and problems should then be identified in a second step, while considering how to reduce GHG emissions by means of such countermeasures. This means linking the existing issues to the potential to reduce GHG emissions in each sector. These two steps in the methodology are to clarify the issues and the barriers to addressing the issues.

TABLE 7.1 Examples of NAMAs submitted by non-Annex 1 Parties

Country	*Emission reduction*	*Mitigation actions with numerical targets*	*Mitigation actions without numerical targets*
Afghanistan	N.A.	–	Preparation of the Initial National Communication (INC), which will include specific mitigation strategies and activities appropriate for the national context
Brazil	36.1–38.9%, by 2020 *No definition of base year*	• Reduction in Amazon deforestation (564 million tons CO_2 (mil tCO_2) • Reduction in "Cerrado" deforestation (104 mil tCO_2) • Restoration of grazing land (83–104 mil tCO_2) • Integrated crop-livestock system (18–22 mil tCO_2) • No-till farming (16–20 mil tCO_2) • Biological N_2 fixation (16–20 mil tCO_2) • Energy efficiency (12–15 mil tCO_2) • Increase the use of biofuels (48–60 mil tCO_2) • Increase energy supply using hydroelectric power plants (79–99 mil tCO_2) • Alternative energy sources (26–33 mil tCO_2) • Iron and steel (replace coal from deforestation with coal from planted forests, 8–10 mil tCO_2)	–
Bhutan	N.A.	–	"Bhutan already sequesters more carbon than is emitted, and we have declared our intent to ensure that our emissions do not exceed our sequestration capacity. This 'Declaration of the Kingdom of Bhutan-Land of Gross Happiness to Save our Planet', signed 11 December 2009 by our Honorable Prime Minister Jigmi Y. Thinley, was also submitted to the President of COP15 in Copenhagen."

continued

TABLE 7.1 Continued

Country	*Emission reduction*	*Mitigation actions with numerical targets*	*Mitigation actions without numerical targets*
China	40–45%, per unit of GDP by 2020, compared to the 2005 level	• Lower CO_2 emissions per unit of GDP by 40–45% by 2020 compared to the 2005 level • Increase the share of non-fossil fuels in primary energy consumption to around 15% by 2020 • Increase forest coverage by 40 million ha and forest stock volume by 1.3 billion m^3 by 2020 from the 2005 levels	–
India	20–25%, per unit of GDP by 2020, compared to the 2005 level	• Agricultural sector emissions will not be included in the emission intensity measurement	–
Indonesia	26%, by 2020 *No definition of base year*	–	1 Sustainable peat land management 2 Reduction in rate of deforestation and land degradation 3 Development of carbon sequestration projects in forestry and agriculture 4 Promotion of energy efficiency 5 Development of alternative and renewable energy sources 6 Reduction in solid and liquid waste 7 Shifting to low emission transportation mode
Korea	30%, by 2020	• Reduce national GHG emissions by 30% from BaU emissions by 2020	–

Maldives	(100%) Carbon neutrality as a country by 2020	• "The Government is presently undertaking detailed work on implementation of this action and will register a request for technological, financial and capacity building support for implementation."	–
Marshall Islands	40%, below 2009 level	• Pursuant to the 2009 National Energy Policy and Energy Action Plan, and subject to the provision of adequate international support	–
Mexico	30%, with respect to the BaU by 2020	• Reduction in total annual emissions of 51 mil tCO_2 by 2012	–
Singapore	16%, below BaU in 2020	–	• "Although a legally binding agreement has yet to be achieved, Singapore will nonetheless begin to implement the mitigation and energy efficiency measures announced under the Sustainable Singapore Blueprint in April 2009." • "When a legally binding global agreement on climate change is reached, Singapore will implement additional measures to achieve the full 16% reduction below BaU in 2020."
South Africa	34%, by 2020 compared to BaU; 42%, by 2025 compared to BaU	–	–

Source: http://unfccc.int/home/items/5265.php (as of June 27, 2010).

What kind of MRV should be established?

The meaning of "MRV" is rather ambiguous, and many discussions are still underway. A clear definition of MRV is required at the beginning stage. Many kinds of MRV have been discussed, such as MRV for actions to reduce GHG emissions, or for the technology, finance, and capacity building to be provided by developed countries. The authors focused on MRV of the GHG emission reduction that is accomplished by NAMA projects, by developing a NAMA planning methodology. In order to implement NAMAs, including target-setting and identifying possible financial resources, it is essential to break down NAMA plans to the level of actual projects. Therefore, Clean Development Mechanism (CDM) methodologies would be good references for the MRV component of NAMA projects, as CDM methodologies were developed for application to project-based activities. This means that the MRV for NAMAs does not need to be established from the beginning, but can be based on an existing MRV system.

How to finance NAMA project implementation?

Financial feasibility is an inevitable concern in implementing NAMAs. Many CDM projects could not be carried out because they lacked practical financial resources. NAMAs will be implemented by a developing country on the premise that bilateral relationships and the financial support of a developed country are essential. Financing from the private sector will also play an important supporting role. Practical financial resources should first be identified for each activity of NAMAs. General discussions, such as regarding the appropriateness of applying for ODA or utilizing private financing for NAMAs in general, will be neither effective nor feasible. The authors focused on assessing the financial feasibility of NAMA projects as an aspect of developing the NAMA planning methodology.

How can the results of an implemented NAMA plan be shared with other developing countries?

The activities and experiences noted above should be shared among developing countries to promote effective NAMA planning throughout a sector. Most developing countries have similar problems in the same sectors, including waste management, transportation, energy efficiency, and so on. Therefore, international conferences and/or training seminars about NAMA and MRV will be good tools for disseminating appropriate methodologies and practical examples.

Methodology for developing NAMA plans

Considering the key issues mentioned above, with the aim of accelerating NAMA plan development by developing countries, the authors have developed a methodology entitled "A Methodology for Developing a NAMA Plan and for Selecting

Priority NAMA Projects: Seven Steps for NAMA Planning," which consists of the seven steps outlined below. An example of an application of this methodology to a study entitled "Feasibility Study for NAMA in Waste and Wastewater Management Sector in Thailand"[6] funded by the Ministry of the Environment, Japan is also provided in what follows.

Step 1: reviewing related policies/plans and their progress in the NAMA sector

In the first step, our objective is to understand and assess related policies/plans and their progress in the NAMA sector by examining existing national policies/plans/targets, and by reviewing the present situation in the sector.

First, we examine existing national policies/plans to find the provisions related to the NAMA sector, such as waste management in the case of the feasibility study, for which the corresponding policies/plans were the "Tenth National Economic and Social Development Plan" and the "Action Plan" of the Pollution Control Department. Next, we extract targets from such policies/plans. In addition, if metropolitan or local governments, such as the Bangkok Metropolitan Authority (BMA) or other cities, have their own policies/plans related to the NAMA sector, we examine them as well, and extract targets from them.

Next, we collect data and information related to the NAMA sector for the past five to ten years in order to understand and assess the progress of the national policies/plans/targets in the NAMA sector. In the feasibility study, the data and information at the national level included past and present waste emission conditions. In addition, if metropolitan or local governments such as the BMA or other cities have detailed data and information on the sector, we also use this data to understand and assess progress made in the sector.

Based on the data and information, we try to understand and/or identify how to manage the existing national policies/plans/targets in order to utilize them to develop an MRV methodology for the NAMA plan (Step 5). We check indicators for the targets, and reporting and verifying modalities and procedures (M/P) to manage them.

In this feasibility study, we divided the waste management sector into four activity areas as shown in Figure 7.1. The results of the exercise in Step 1 are shown in Table 7.2.

Step 2: setting reference scenarios in the NAMA sector

In this next step, we set reference scenarios (baseline scenarios) based on an analysis of barriers facing existing policies/plans in the NAMA sector.

We identify barriers to implementing existing policies/plans/targets in the NAMA sector, taking into consideration the existing policies/plans/targets and related progress studied in the previous step. Possible barriers may include financial, technical, human resource related, and prevailing practice related barriers.

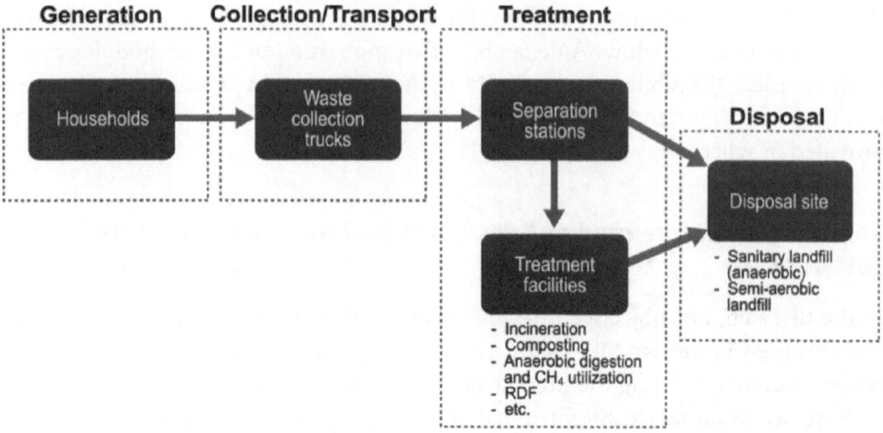

FIGURE 7.1 Waste management sector activities (source: author).

Next, we make a prediction of the future trends (after ten to 20 years) in the barriers identified above, and establish reference scenarios in the sector based on the results of these predictions. If the implementation of an existing policy/plan is going well, the reference scenario and the targets may be almost the same. If implementation is very slow due to critical barriers, the reference scenario may be almost the same as "business as usual" (BaU). In this feasibility study, for example, we tried to set reference scenarios for all the activities we identified in the sector: generation, collection/transport, treatment, and disposal.

Finally, we consider possible countermeasures to address the barriers in the sector we identified in the sub-steps above. In the feasibility study, these countermeasures may include institutional countermeasures such as increasing appropriate budget allocations for waste management systems in local communities, educational countermeasures such as encouraging and enhancing awareness-raising for the public on solid waste separation, and technical countermeasures such as introducing and/or increasing composting systems. The results of the exercise in Step 2 are shown in Table 7.3.

Step 3: setting NAMA targets in the NAMA sector

In Step 3, the objective is to set NAMA targets after identifying the main sources of GHG emissions and actual GHG mitigation measures in the NAMA sector.

First, we use the 2006 IPCC Guidelines for National Greenhouse Gas Inventories and other appropriate references to identify the main GHG emission sources in the NAMA sector. In the feasibility study, for example, we can identify CO_2 emissions from MSW collecting trucks, and CH_4 emissions from landfill sites, as sources of GHG emissions.

According to the main sources of GHG emissions thus identified, we select different types of actual GHG mitigation measures for the NAMA sector. The measures can be classified into the following three categories:

TABLE 7.2 Step 1: reviewing related policies/plans and their progress in NAMA sector

Waste management activities	*Step 1*			
	Existing policy/plan	*Targets*	*Policy/plan progress*	*M/P of the policy/plan*
Generation	• Abhisit Vejjajiva's strategy formation, Policy Statement No. 5, Policy on Land, Natural Resources, and the Environment [NESDB 2009] • The Tenth National Economic and Social Development Plan 2007–2011 [The Tenth] • National Strategic Plan for Municipal General and Hazardous Solid Waste Management [ONEP 2005]	• Limit the production of waste in urban areas to no more than 1 kg/person/day [The Tenth, Chapter 1] • Key Performance Indicators (KPIs) from 2005 to 2014 [ONEP 2005] • Bangkok, Pattaya City and other large municipalities: maximum of 1 kg/p/d • Muang municipalities: 0.8 kg/p/day • Tambon municipalities: 0.6 kg/p/d • Other local authorities: 0.4 kg/p/d • Total community-generated solid waste: not to exceed 14 million tons/year	• National average amount of waste production is 0.64 kg/p/d in 2008 [PCD 2008] • However, some areas, especially in urban areas, produce more than 1 kg/p/d. Therefore, the target should be 1 kg to prevent an increase in waste, alongside promotion of recycling through the 3Rs (reuse, reduce, recycle)	• TBD • TBD • TBD

continued

TABLE 7.2 Continued

Waste management activities	*Step 1*			
	Existing policy/plan	*Targets*	*Policy/plan progress*	*M/P of the policy/plan*
Collection/transport	• The Policy for Preventing and Eliminating Pollution under the Enhancement and Conservation of National Environmental Quality Act, 1997–2016 [PCD 1997])	• Not less than 90% of total waste in each local area (including from large markets, etc.) shall be collected, transported, and disposed of [ONEP 2005] • 100% of municipal solid waste (MSW) can be collected in a municipality [PCD 1997] • No more than 10% can remain in an outdoor area [PCD 1997]	• 84% of MSW was collected from municipality services and 16% remained uncollected [PCD 2010] • Hazardous waste is still mixed with municipal waste; there are only 21 collection sites for hazardous waste in all of Thailand	• TBD • TBD • TBD
Treatment		• Reuse at least 30% of the waste throughout the country [The Tenth, Chapter 5] • At least 40% of total generated waste (5.6 million t/y) will be segregated and recycled [ONEP 2005] • Reuse at least 15% of waste throughout the country [PCD 1997] • The capacity of an established waste to energy (WTE) project must not be less than 100 MW, and at least one facility will be developed as a WTE pilot model [PCD 2010]	• Treatment method for MSW is focusing on recycling and composting; in 2009 there was a waste recovery potential of approximately 25% • Industrial waste recycling was approx. 68% • Only 38% of solid waste was treated and disposed of properly	• TBD • TBD • TBD

Disposal	• Ensure proper disposal of at least 80% of all hazardous waste from communities and industries [Chapter 1, The 10th] • The residue disposed of after total waste segregation and recycling will not exceed 8.4 million t/y [ONEP 2005] • The capacity of an established WTE project must not be less than 100 MW, and at least one facility will be developed as a WTE pilot model [PCD 2010]	• Municipal waste: approx. 47% of collected waste was disposed of using sanitary disposal methods, and approx. 53% using improper disposal methods • In 2010 there were 107 sanitary landfill sites, three incineration sites, and four integrated system sites	• TBD • TBD • TBD

Sources:
NESDB 2009: Abhisit Vejjajiva's Strategy Formation, the fifth policy, on Land, Natural Resource and Environment.
The Tenth: The Tenth National Economic and Social Development Plan 2007–2011, Office of the National Economic and Social Development Board.
PCD 2009: Thailand Pollution Situation in 2009, Pollution Control Department.
PCD 1997: The Policy for Preventing and Eliminating Pollution under the Enhancement and Conservation of National Environmental Quality Act, 1997–2016.
ONEP 2005: National Strategic Plan for Municipal General and Hazardous Solid Waste Management, the Office of Natural Resources and Environmental Policy and Planning.
PCD January 2011: Draft concept and direction, pollution management plan 2012–2016 (in Thai).
PCD 2010: Presentation paper, Overview on Municipal Solid Waste Management in Thailand, November 2010.
TBD: to be determined.

TABLE 7.3 Step 2: setting reference scenarios in the NAMA sector

Waste management activities	*Step 2*			
	Barriers identified	*Predicted condition of barriers*	*Reference scenarios*	*Countermeasures to address barriers*
Generation	• Some consumers prefer non-renewable products, and do not trust the quality of recycled products [ONEP 2005] • It is inconvenient to find products made from recycled materials [ONEP 2005T] • People do not cooperate effectively in separating household solid waste [ONEP 2005T]	• As it is not easy to change consumer preferences, the barrier will exist in the long term • As it is not easy to expand the market for recycled materials, the barrier will exist in the long term • As it is not easy to change human behavior, the barrier will exist in the long term	• The amount of MSW increases according to population increase and economic growth • The rate of increase is higher in local communities • Consumer preferences/behavior do not change dramatically • Therefore, the target is attainable in larger cities, but may not be easy to attain in local communities	• *Bring in regulations requiring manufacturers and the public to segregate waste at source [ONEP 2005] (I)* • Improve capacity of Local Administration. Organizations (LAOs) concerning waste reduction, reuse, and recycling (E) • *Increase appropriate budget allocation for waste management systems in local communities (I)* • Encourage and enhance raising public awareness of solid waste separation [ONEP 2005] (E) • Improve education scheme concerning consumer packaging for recycling or deposit refund schemes [ONEP 2005] (E)

Collection/transport	• Waste collection and transport are not available everywhere, especially in small communities [ONEP 2005] • The waste collection and transport systems make source segregation of waste difficult, and are operated in ways that cause disturbances [ONEP 2005] • LAOs do not properly allocate budgets for collecting and transporting waste [ONEP 2005] • Waste collection routes are often repetitive and complicated, and in some communities access routes for collection are too narrow or parking is prohibited [ONEP 2005] • Waste collection vehicle maintenance budgets are inadequate and there are not enough vehicles, either overall or in good condition for use [ONEP 2005]	• Improvement in the collection and transport system, including LAOs, can be expected in major cities in the relatively short term. However, this will be difficult to attain in small communities. Therefore, the barrier will exist in the long term in small communities • Local governments do not clearly allocate budgets for solid waste management and system design [ONEP 2010]. Therefore, the barrier will exist in the long term • Government and private investments for comprehensive solid waste management are limited and there is also inexplicit investment coordination between government and the private sector [ONEP 2010]. Therefore, the barrier will exist in the long term	• Collection and transportation system will remain in an inadequate condition, especially in small communities • Therefore, the target is attainable in larger cities, but may not be easy to attain in local communities • Waste collection route conditions will remain complicated and inefficient • Narrow access and limited parking will not improve • The number of collection vehicles will increase due to population growth and increasing waste. However, the condition of the vehicles will not improve dramatically, particularly for vehicles in small communities • Condition of the fee and tax systems will remain inefficient	• *Specify solid waste disposal fees that reflect actual costs, and set up new service rates for solid waste collection [ONEP 2005] (I)* • *Develop criteria and a toolkit for waste collection and transport for LAOs, and introduce regulations to control private sector waste collection and disposal [ONEP 2005] (I)* • Improve capacity of LAOs concerning effective waste collection schemes (E) • *Increase appropriate budget allocation for waste management systems in local communities (I)* • Introduce highly efficient collection vehicles appropriate for the amount of waste and road conditions (such as narrow access) in local communities (T)

continued

TABLE 7.3 Continued

Waste management activities	*Step 2*			
	Barriers identified	*Predicted condition of barriers*	*Reference scenarios*	*Countermeasures to address barriers*
	• Fee collection is ineffective and cannot reflect the actual capital cost [ONEP 2010] • Tax tariffs from contractors, taxes, and local collection fees are not successful and cannot reflect the actual cost [ONEP 2005T]	• As there are many conflicts of interest among stakeholders, improvement of fee/tax related issues is not easy, and the barrier will exist in the long term	• Private sector business opportunities will be limited due to the inadequate fee and tariff system	
Treatment	• The built-up system is located in unsuitable places (too close to local communities) [ONEP 2010] • Waste treatment system maintenance knowledge and budgets are lacking • Lack of knowledge, technology, and/or capital funding to support and promote manufacturing production from secondary resources	• Local governments do not clearly allocate a budget for solid waste management and system design [ONEP 2010]. Therefore, the barrier will exist in the long term • Currently two of the three existing incinerators are dysfunctional [PCD 2010] •	• It will be difficult to install incineration facilities • The number of other treatment systems installed, such as composting systems, will be limited and will not increase dramatically • Manufacturers will be reluctant to use secondary resources and demand for secondary resources will remain low	• Introduce and/or increase incineration facilities of an adequate scale at appropriate sites (T) • Introduce and/or increase composting systems (T) • Introduce/increase anaerobic treatment systems (T) • Consider the feasibility of introducing refuse-derived fuel (RDF) systems (T) • Improve capacity of LAOs concerning effective waste collection schemes (E)

	• Manufacturers prefer to use primary resources, due to the low quality of domestic secondary resources, such as their high impurity content • Manufacturers need to develop and adjust the production processes to use secondary resources and it raises costs • Public cooperation is lacking and public participation is inadequate in waste segregation at source [ONEP 2005]	As it is not easy to change consumer preferences, and decision makers in private companies often prefer a short-term payback period, they will be reluctant to invest in modifying their production processes to make use of secondary resources, which do not easily increase. Therefore, the barrier will exist in the long term • Glass, paper, and plastic are relatively easy to reuse and recycle but food waste is difficult [ONEP 2005] • As it is not easy to change human behavior, the barrier will exist in the long term	• Some kinds of waste such as glass, plastic, and paper will be reused and recycled, but food waste will remain unused and the quantity will increase • Therefore, the targets for 3R are attainable for the latter materials, but may not be easy to attain for food waste	• *Increase appropriate budget allocation for waste management systems in local communities (I)* • *Set up standards for waste treatment operations [ONEP 2005] (I)* • *Introduce regulations for the use of secondary resources and/or provide tax or investment incentives for manufacturers where they can replace primary resources [ONEP 2005] (I)* • *Require government units to purchase products made from recycled and/or environmentally friendly materials [ONEP 2005] (I)*

continued

TABLE 7.3 Continued

Waste management activities	*Step 2*			
	Barriers identified	*Predicted condition of barriers*	*Reference scenarios*	*Countermeasures to address barriers*
Disposal	• People disagree with and oppose solid waste disposal facilities, because their locations are close to communities or natural water resources [ONEP 2010] • The number of operating landfills is inadequate • There are insufficient sanitary landfills, especially those offering full disposal operations • It is very difficult to determine a location for sanitary landfills, although suitable sites exist, because of robust opposition by local populations • When landfills are filled and closed, there is no effort to control their long-term environmental impact	• As it is not easy to change people's opinions from disagreement to agreement and understanding with regard to solid waste disposal facilities, the barrier will exist in the long term • Food waste is difficult to reuse and recycle [ONEP 2005]. It will cause more demand for landfill sites, but for the reasons mentioned in the left-hand column, the barrier will exist in the long term • Inappropriate control of closed landfill sites causes more opposition from local communities, introducing a vicious circle; thus the barrier will exist in the long term	• The number of landfill sites will increase, but most of them will be small and medium scale • Landfill sites in operation will be maintained properly, but capacity will overflow, especially in small cities, according to population and economic growth • Therefore, the target is attainable in larger cities, but may not be easy to attain in local communities • It is difficult to introduce anaerobic treatment systems in closed sites • The conditions of fee and tax systems remain inefficient	• Introduce and/or increase sanitary landfill sites (T) • Introduce and/or increase semi-aerobic treatment systems, as a measure to reduce environmental pollution (T) • Improve capacity of LAOs concerning effective waste collection schemes (E) • *Increase appropriate budget allocations for waste management systems in local communities (I)* • *Introduce regulations for landfill monitoring, management, and post-closure control (I)* • *Set up standards for waste disposal operations, and a system of waste disposal fees that matches actual operating costs [ONEP 2005] (I)*

	• Disposal fees do not cover landfill operating and maintenance costs • Waste disposal system maintenance knowledge and budgets are lacking • Laws concerning general and hazardous municipal solid waste disposal are inconsistent, and cooperation and coordination among offices are lacking [ONEP 2005]	• As there are many conflicts of interest among stakeholders, improvement of fee/tax related issues is not easy and the barrier will exist in the long term • Local governments do not clearly allocate a budget for solid waste management and system design [ONEP 2010]. Therefore, the barrier will exist in the long term	• Private sector business opportunities will be limited due to the inadequate fee and tariff systems • Policies, strategies and plans will be promoted but practical solutions will be limited	• Provide technical and management backup knowledge for LAOs [ONEP 2005] (E)

Sources:
ONEP 2010: Chapter 4, Publication on Technology of Solid Waste Management Disposal, the Office of Natural Resources and Environmental Policy and Planning.
ONEP 2005: Executive Summary: National Strategic Plan for Municipal General and Hazardous Solid Waste Management, the Office of Natural Resources and Environmental Policy and Planning (PDF file created in 2005).
ONEP 2005T: [Thai Version] Background and Justification, National Strategic Plan for Municipal General and Hazardous Solid Waste Management, the Office of Natural Resources and Environmental Policy and Planning.

Note
Legend of countermeasures: I: Institutional, E: Educational, T: Technical

i. implementation of countermeasures against an identified barrier that directly reduce GHG emissions (quantitative reduction, e.g., municipal waste composting);
ii. introduction of new and/or advanced technology that directly reduces GHG emissions (quantitative reduction, e.g., semi-aerobic landfill systems); and
iii. implementation of countermeasures against an identified barrier that indirectly reduce GHG emissions (qualitative contribution, e.g., increasing appropriate budget allocations for waste management systems in local communities, awareness-raising, education).

Next, we roughly estimate the potential for GHG emission reductions with the actual and quantitative GHG mitigation measures identified in the NAMA sector within the next ten-year period. According to the results, we set NAMA targets considering the potential to attain the actual GHG mitigation measures identified in the NAMA sector. In the feasibility study, for example, we set NAMA targets only for the treatment and disposal activities. The results of the exercise in Step 3 are shown in Table 7.4.

Step 4: developing priority NAMA projects in the NAMA sector

Using a selection methodology, in this step our objective is to develop priority NAMA projects.

Our selection methodology employs a scoring formula such as that in the evaluation Table 7.5. Evaluation items are selected to take account of the urgency, advantages, and benefits of the projects. Each item is scored from one (low) to five (high). In addition, each score is weighted according to an expert assessment of the importance of the item. Comparing the total scores, we can then prioritize the NAMA projects (Table 7.5).

In particular, collecting priority NAMA project candidates includes cooperation with local counterparts: we inform resource persons and stakeholders all over the country of the selection process, especially in the capital city and other major municipalities. In the feasibility study, we discussed the candidate selection with the Thailand Greenhouse Gas Management Organization (TGO) and BMA members of the Host Country Committee. We next selected priority NAMA projects in the NAMA sector using the selection methodology noted above, and TGO/BMA confirmed their appropriateness.

Step 5: establishing MRV methodologies for the NAMA plan

In Step 5, the objective is to study and establish MRV methodologies for the NAMA targets and priority NAMA projects under the NAMA plan. Progress management methodologies (which are assumed to be MRV methodologies) are established for comprehensive targets in the sector (e.g., the annual amount of waste per capita) and MRV methodologies are set up for the priority NAMA projects. As the

TABLE 7.4 Step 3: setting NAMA targets in the NAMA sector

Waste management activities	*Step 3*			
	Main GHG emission sources identified	*Actual GHG mitigation measures*	*GHG emission reduction potential*	*NAMA target*
Generation	• *CH_4 emissions from landfill sites can be reduced by countermeasures in this stage*	• 3R (reduce, reuse, recycle) promotion campaign in BMA	• To be estimated	• To be determined
		• 3R promotion campaign in Phuket	• To be estimated	• To be determined
		• 3R promotion campaign in small- and medium-size municipalities	• To be estimated	• To be determined
Collection/transport	• CO_2 emissions from MSW collection trucks	• Introduction of compressed natural gas (CNG) trucks for MSW collection	• About 20% CO_2 emission reduction compared to diesel fueled trucks	• To be determined
		• Optimization of truck sizes for MSW collection, and of collection routes	• To be estimated	• To be estimated
		• Utilization of bio-fuels by MSW collection trucks	• To be estimated	• To be estimated

continued

TABLE 7.4 Contnued

Waste management activities	*Step 3*			
	Main GHG emission sources identified	*Actual GHG mitigation measures*	*GHG emission reduction potential*	*NAMA target*
Treatment	• *CH_4 emissions from landfill sites can be reduced by countermeasures in this stage*	• Introduction of incineration plants in major municipalities including BMA • Introduction of composting plants in BMA • Introduction of composting facilities in small- and medium-size municipalities • Introduction of anaerobic digestion plants in major municipalities including BMA • Introduction of RDF production facilities in BMA	• 3,000,000 tCO_2/year (introduce 30 plants) • 1,000,000 tCO_2/year (all BMA MSW) • 3,500,000 tCO_2/year (all MSW except BMA) • 1,050,000 tCO_2/year (introduce 30 plants) • 900,000 tCO_2/year (introduce 30 plants)	• One incineration plant in BMA by 2020 • Five composting plants in BMA by 2020 • Ten composting plants in major municipalities by 2020 • Ten anaerobic digestion plants in major municipalities by 2020 • One RDF production facility in BMA by 2020 *Targeted emission reduction: tCO_2/year*

Disposal	• CH_4 emissions from landfill sites	• Introduction of landfill gas (LFG) recovery systems and power generation in closed major sanitary landfills	• 500,000 tCO_2/year (introduce 10 plants)	• Two LFG recovery system and power generation in closed major sanitary landfills by 2020
		• Introduction of semi-aerobic systems in closed and operating sanitary landfills	• 4,800,000 tCO_2/10 year (emission reduction, 2011–2020)	• 20 semi-aerobic system in closed and operating sanitary landfills by 2020
		• New construction of semi-aerobic landfill systems in small- and medium-size municipalities	• 2,300,000 tCO_2/year	• 30 new construction of semi-aerobic landfills system in small- and medium-size municipalities by 2020
		• New construction of semi-aerobic landfill systems in BMA		• Five new construction of semi-aerobic landfills system in BMA by 2020 *Targeted emission reduction: tCO_2/year*

TABLE 7.5 Selection methodology for priority NAMA projects (example)

Evaluation items (tentative)	*Score*					*Weighting*	*Total*
	1	*2*	*3*	*4*	*5*		
1 Urgency			○			×2	6
2 GHG emission reduction					○	×2	10
3 Technological advantage					○	×2	10
4 Co-benefit effect					○	×1	5
5 Cost–benefit			○			×1	3
Total score	*34*						

Source: authors.

former activity includes obtaining information from data collection according to verification modalities and procedures, we pursue development of the methodology that was considered in Step 1: the management methodology for national policies/plans/targets. For the latter, we establish simplified MRV methodologies for actual projects, referring to approved CDM methodologies. In the feasibility study, we developed tentative MRV methodologies according to the CDM for the latter, and the former continue to be under discussion.

Step 6: identifying possible financial resources for implementing the NAMA plan

In Step 6, we identify sources of financing for the NAMA plan, including priority NAMA projects. Possible sources of funding for priority NAMA projects may include: (i) funds provided by the host country's government and/or private entities; (ii) support from developed countries and/or international financial organizations; and (iii) cooperation between private entities in the host country and from outside the country. Combining these options, we identify an appropriate utilization of financial resources for the sector and subsectors. In addition, the amount of GHG emission reduction compared to the reference scenario may be supplied to the Bilateral Offset Credit mechanisms being considered by the Japanese government.

In the feasibility study, for the priority NAMA project candidates in the treatment and disposal activity areas, we proposed utilizing a public–private partnership/private finance initiative (PPP/PFI) financed by JBIC GREEN[7] under the bilateral credit mechanism between Japan and Thailand as one of the highly feasible financial resources (italicized in Table 7.6).

Step 7: determining an appropriate schedule for implementing the NAMA plan

The objective of the seventh and final step is to settle the schedule for implementing the priority NAMA projects under the NAMA plan. Based on the financial

TABLE 7.6 Steps 4, 5, 6, and 7

Waste management activities	*Step 4*		*Step 5*	*Step 6*	*Step 7*
	Priority NAMA project candidates	*Priority score*	*MRV Methodologies*	*Financial resources*	*Possible schedule*
Generation	• To be considered		• To be considered	• To be considered	• To be considered
Collection/ Transport	• Introduction of CNG trucks for MSW collection in BMA	55	• CDM: AMS-III.C, III.S	• Budget of Thai Government and financing from Japanese private companies under the bilateral credit mechanism (tentative) between Japan and Thailand	• To be decided
	• Introduction of large-size trucks for MSW collection in BMA	60	• CDM: AMS-III.C, III.S		
	• Utilization of bio-fuels by MSW collecting trucks in BMA	70	• CDM: AMS-III.T, III. AK		
Treatment	• Introduction of MSW incineration plant in BMA	68	• CDM: AM0025,	• *Utilization of PPP/PFI financed by JBIC GREEN under the bilateral credit mechanism (tentative) between Japan and Thailand*	• To be decided
	• Introduction of MSW composting plants in BMA	80	• CDM: ACM-III.F		
	• Introduction of night-soil composting plants in BMA	88	• CDM: ACM-III.F		
	• Introduction of composting facilities in Chiang Mai	85	• CDM: ACM-III.F		
	• Introduction of composting facilities in Chanthaburi	86	• CDM: ACM-III.F		
	• Introduction of anaerobic digestion plants in Phuket	62	• CDM: ACM-III.AO		
	• Introduction of RDF production facilities in Ayutthaya	71	• CDM: ACM-III.E		

continued

TABLE 7.6 Continued

Waste management activities	*Step 4*		*Step 5*	*Step 6*	*Step 7*
	Priority NAMA project candidates	*Priority score*	*MRV Methodologies*	*Financial resources*	*Possible schedule*
Disposal	• Introduction of LFG recovery system and power generation in closed major sanitary landfills in Nonthaburi	79	• CDM: ACM 0001, AMS-III.G	• *Utilization of PPP/PFI financed by JBIC GREEN under the bilateral credit mechanism (tentative) between Japan and Thailand*	• To be decided
	• Introduction of semi-aerobic system in closed and operating sanitary landfills in Nakhon Si Thammarat.	86	• CDM: NM0333, ACM 0001, AMS-III.G; IPCC		
	• Introduction of semi-aerobic system in closed and operating sanitary landfills in Songkhla	85	• CDM: NM0333, ACM 0001, AMS-III.G; IPCC		
	• New construction of semi-aerobic landfills system in BMA	82	• CDM: NM0333, ACM 0001, AMS-III.G; IPCC		
	• New construction of semi-aerobic landfills system in Chon Buri	80	• CDM: NM0333, ACM 0001, AMS-III.G; IPCC		

plans noted above for the priority NAMA projects, we work out an appropriate schedule for implementing these projects, taking into consideration the feasibility of the projects. In the feasibility study, the schedule continues to be under discussion. The results of the exercises in Step 4 to Step 7 are shown in Table 7.6.

Toward developing NAMAs that enable MRV

As mentioned in the previous section, wide-ranging discussions about NAMAs are still ongoing. These include technical issues (e.g., setting reference scenarios (Asselt *et al.* 2010; Niederberger and Kimble 2011)), financial issues (e.g., financing for NAMA projects (Bird *et al.* 2011; Cheng 2010; Neuhoff *et al.* 2009)), market related issues (e.g., crediting of NAMA projects (Olsen *et al.* 2009; Okubo *et al.* 2011)), and so on. Among these, only the importance of contributing to sustainable development was confirmed in this study. For this purpose, developing countries expect technical and financial support from developed countries, not only for GHG emission reduction, but also for social and economic development and poverty eradication.

In this study, the proposed methodology was successful in identifying current issues and problems in the targeted sector, identifying barriers to addressing these issues, and identifying countermeasures whereby GHG emissions could be reduced in the sector. Such technical countermeasures have the potential to be NAMA projects. The methodology can also identify financial resources including bilateral support, as well as devising an implementation schedule for NAMA projects, which can quantify GHG emission reduction.

However, it was difficult to develop NAMA projects using the methodology for institutional and educational activities that have an indirect effect on reducing GHG emissions, such as 3R (reduce, reuse, recycle) initiatives and awareness-raising. Table 7.7 shows a summary of the countermeasures selected, from institutional, educational, and technical points of view. Among these, only technical countermeasures have quantifiable effects on reducing GHG emissions.

Because of the difficulty in quantifying the GHG emission reduction effects of institutional and educational countermeasures, no methodology was developed to measure, report, and verify such effects either under the CDM or in other similar GHG related mechanisms, and only limited examples are available as references (JICA 2009). If we can quantify these indirect effects, it will enable MRV for any kind of NAMA. In order to develop a methodology to quantify these indirect effects, many assumptions would be required, and many default values should be prepared according to good practices already implemented by developed countries. For example, 3R initiatives are an example of educational countermeasures in the waste management sector that have already been carried out by many developed countries, including Japan. It will be very useful if such indirect efforts to reduce GHG emissions can be quantified using the results of these initiatives in developed countries. The authors consider that the quantification of these indirect GHG emission reduction effects, such as institutional and educational countermeasures, is the next step in elaborating a methodology for developing practical NAMAs that enable MRV.

TABLE 7.7 Summary of selected countermeasures

Generation	*Collection/transport*	*Treatment*	*Disposal*
Institutional countermeasures			
• Increase appropriate **budget** allocations for waste management systems in local communities • Introduce **regulations** requiring manufacturers and the public to segregate waste at source	• Increase appropriate **budget** allocations for waste management systems in local communities • Introduce **regulations** to control private sector waste collection and disposal • **Specify fees** for solid waste disposal • Set up **new service rates** for solid waste collection • **Develop criteria** and a toolkit for waste collection and transport for LAOs	• Increase appropriate **budget** allocations for waste management systems in local communities • Introduce **regulations** for the use of secondary resources • Set up **standards** for waste treatment operations • **Provide tax incentives** or **investment incentives** for manufacturers if they can replace primary resources • **Require** government units to purchase products made from recycled and/or environmentally friendly materials	• Increase appropriate **budget** allocations for waste management systems in local communities • Introduce **regulations** for landfill monitoring, management, and post closure control • Set up **standards** for waste disposal operations, and a system of waste disposal fees that matches actual operating costs

Educational countermeasures			
• **Improve** capacity of LAOs concerning waste reduction, reuse, and recycling • Encourage and enhance **raising public awareness** of solid waste separation • **Improve** education schemes concerning consumer packaging for recycling or deposit refund schemes	• **Improve** capacity of LAOs concerning effective waste collection schemes	• **Improve** capacity of LAOs concerning effective waste collection schemes	• **Improve** capacity of LAOs concerning effective waste collection schemes • **Provide** technical and management backup **knowledge** for LAOs
Technical countermeasures			
–	• Introduce highly efficient **collection vehicles** appropriate for the amount of waste and road conditions (such as narrow access) in local communities	• Introduce and/or increase **incineration facilities** of an adequate scale at appropriate sites • Introduce and/or increase **composting systems** • Introduce/increase **anaerobic treatment systems** • Consider the feasibility of introducing **RDF systems**	• Introduce and/or increase **sanitary landfill** sites • Introduce and/or increase **semi-aerobic treatment systems**, as a measure to reduce environmental pollution

Source: authors.

Note
Bold indicates the keyword for each countermeasure.

Notes

1 We usually employ the term "MRV" with "M" as *measurement*, "R" as *reporting*, and "V" as *verification*.
2 These actions are called "supported NAMAs."
3 These actions are called "domestic NAMAs."
4 The Copenhagen Accord (FCCC/CP/2009/L.7) stipulates

> Non-Annex I Parties to the Convention will implement mitigation actions, including those to be submitted to the secretariat by non-Annex I Parties in the format given in Appendix II by 31 January 2010, for compilation in an information document, consistent with Article 4.1 and Article 4.7 and in the context of sustainable development.

5 Personal communication at a closed meeting in March 2011.
6 PC-Institute for Global Environment Research (PC-iGER), "Feasibility Study for NAMA in Waste and Wastewater Management Sector in Thailand," Global Environment Research Foundation (GEC) funded by the Ministry of the Environment, Japan (2011).
7 Global action for Reconciling Economic growth and ENvironmental preservation (GREEN): JBIC uses this loan scheme to support projects in developing countries that are having a favorable impact on GHG emission reduction. Source: www.jbic.go.jp/en/about/news/2010/0427–01/index.html.

References

Asselt van, H., Berseus, J., Gupta, J., and Haug, C. (2010) "Nationally appropriate mitigation actions (NAMAs) in developing countries: Challenges and opportunities," *Climate Change: Scientific Assessment and Policy Analysis*, Bilthoven: Netherlands Environmental Assessment Agency PBL, Report No. 500102035, University of Amsterdam.

Bird, N., Brown, J., and Schalatek, L. (2011) "Design challenges for the Green Climate Fund," *Climate Finance Policy Brief*, 4, January. Available at: www.boell.org/downloads/Boell-ODI_Design_Challenges_Green_Climate_Fund.pdf (accessed November 18, 2011).

Cheng, C-C. (2010) "A new NAMA framework for dispersed energy end-use sectors," *Energy Policy*, 38(10), 5614–24.

JICA (Japan International Cooperation Agency) (2009) *Project Study on Quantification of Greenhouse Gases (GHGs) Emission Reduction*, Tokyo: JICA.

Okubo, Y., Hayashi, D., and Michaelowa, A. (2011) "NAMA crediting: How to assess offsets from and additionality of policy-based mitigation actions in developing countries," *Greenhouse Gas Measurement and Management*, 1(1), 37–46.

Olsen, K.H., Fenhann, J., and Hinostroza, M. (Eds.) (2009) *NAMAs and the Carbon Market, Perspectives Series*, Roskilde: UNEP Risø Centre CD4CDM.

Neuhoff, K., Fankhauser, S., Guerin, E., Hourcade, C.J., Jackson, H., Rajan, R., and Ward, J. (2009) *Structuring International Financial Support to Support Domestic Climate Change Mitigation in Developing Countries*, Cambridge, UK: Climate Strategies.

Niederberger, A.A. and Kimble, M. (2011) "MRV under the UN climate regime: Paper tiger or catalyst for continual improvement?" *Greenhouse Gas Measurement and Management*, 1(1), 47–54.

UNFCCC (2007) Decision -/CP.13, Bali Action Plan, p. 19.

UNFCCC (2010) Draft Decision -/CP.16, Outcome of the Work of the Ad Hoc Working Group on Long-term Cooperative Action under the Convention, p. 29.

8

EFFECTIVENESS OF DEVELOPMENT POLICY LENDING IN MAINSTREAMING AND FINANCING CLIMATE CHANGE

Case studies from Indonesia and Vietnam[1]

Taro Katsurai[2] and Yuka Murakami[3]

Introduction

In December 2010, the 16th session of the Conference of Parties (COP16) to the United Nations Framework Convention on Climate Change (UNFCCC) was held in Cancun, Mexico, and adopted the Cancun Agreements. The Cancun Agreements, once again, called for global efforts to "reducing global greenhouse gas emissions so as to hold the increase in global average temperature below 2 °C above preindustrial levels." The 2 °C warming from the pre-industrial level is said to be the level in which the world can expect to experience "dangerous" consequences. Growing scientific evidence reveals that maintaining the warming within this threshold is difficult (Parry, 2009). Furthermore, predictions of the consequences of even a 2 °C increase are becoming increasingly severe (Smith *et al.*, 2009).[4] These current understandings clearly suggest the need for two measures. First, worldwide mitigation efforts must include developing countries. Even though their current contribution of global GHG emissions is still small, it is, nevertheless, critical to take actions and considerations immediately in order for them to avoid locking into a high carbon economy (Shalizi and Lecocq, 2009). Second, because severe climate change impacts are unavoidable, adaptation efforts to prepare for such events are also paramount (Parry *et al.*, 2009).

Does this lead to the conclusion that developing countries should abandon their current development strategies and completely reorganize them around climate change? Probably not because as pointed out by Martin Parry, former co-chair of IPCC Working Group II, future vulnerability[5] depends more on the level and quality of development than on climate change (Parry, 2009). In other words, quality development or sustainable development is an effective, if not the most effective, climate change response (Parry, 2009). Developing countries, therefore, should continue pursuing their current sustainable development strategies;[6] the difference is that it has to be implemented better and their concept of sustainability to include considerations for climate change.

Development cooperation agencies have already expressed their full commitment to assisting developing countries pursue this approach. Agencies are rapidly increasing both the quantity and scale of climate change adaptation and mitigation projects as well as mainstreaming climate change[7] into all their projects. In spite of this, however, there are two apparent shortcomings. Although donor-assisted projects are being climate-mainstreamed, non-donor-assisted projects may remain climate-unconsidered. Thus, the first challenge will be to find a means through which agencies' efforts can reach all development efforts especially the ones not financed by them but financed domestically or privately. Meeting this first challenge, however, also leads to a second one: how can development cooperation agencies channel their financing to cover the additional cost entailed by mainstreaming climate change or integrating mitigation and adaptation measures into development projects, especially if such projects are financed domestically?

We argue here that an ODA financing modality called development policy lending (DPL) is an effective means of addressing both these challenges. In the following section, we provide a general overview of DPL, followed by reviews of climate change DPLs in Indonesia and Vietnam. In the final section, we present the reasoning behind our conclusions that DPL is an effective mechanism for addressing these challenges.

Overview of development policy lending

DPL is a type of policy-based lending instrument in which financing is provided as direct support to the recipient government's budget and is based on a set of policy or institutional reforms to be undertaken by the recipient (World Bank, 2006a). DPL came into shape in response to shortcomings of the previous policy-based lending instrument, which was called structural adjustment lending or SAL (World Bank, 2006b). Let us first review SAL in order to delineate DPL.[8]

After the two oil shocks in the 1970s and the subsequent world recession, many developing countries faced severe balance of payment problems. These issues were considered the result of structural problems rather than of short-term shocks, and therefore something beyond short-term assistance was needed to improve the balance of payment or foreign currency reserve. Thus the World Bank Board approved a new financing instrument, the SAL, in February 1980. Contrary to traditional project-type assistance in which financing is made against actual progress of an investment project, SAL provided financing over a period of several years in return for reforms that were intended to stabilize the macro-economy while maintaining growth. The set of reforms varied from country to country but in most cases included devaluation of national currencies, liberalization of trade policies, reduction in government spending, enhancement of price incentives, reduction of subsidies, and privatization and liberalization of internal markets and institutions.[9]

SAL was greatly criticized during its implementation in the 1980s and 1990s; the criticisms can be summarized as follows.[10] First, SAL exacerbated the situation of the poor since reduction in government spending often impacted programs intended to help the poor. Second, SAL over-emphasized market mechanisms and underplayed the potential role of government in areas such as trade protection, subsidized credit, and industrial policies. Third, SAL imposed strict short-term macroeconomic stabilization measures that sacrificed long-term development. Last, SAL sustained misguided policies and mal-motivated governments instead of letting them run their natural course (Summers and Pritchett, 1993).

Over the decades, efforts to overcome these criticisms were made both from the operation and research sides[11] which eventually resulted in the establishment of development policy lending, in 2004. DPL departs from SAL in the following ways (World Bank, 2006b). First, the focus shifted from macroeconomic stabilization and adjustment to poverty reduction and growth. The World Bank's current Operational Policy stipulates that "development policy lending aims to help a borrower achieve sustainable reductions in poverty through a program of policy and institutional actions that promote growth and enhance the well-being and increase the incomes of poor people."[12] This differs from the earlier Operational Directive of SAL which stated that the World Bank adjustment lending was introduced "to provide temporary balance-of-payments financing to member countries while stabilization and adjustment measures took effect."[13] Second, SAL offered prescription whereas DPL stressed the importance of procedures for developing the policy reform program. This can be clearly seen in the Operational Manual. SAL manual prescribes market deregulation, trade liberalization, and other policies. In contrast, DPL was careful not to prescribe policy but only to describe the principles and procedures, such as participation, coordination, and harmonization (World Bank, 2006b).

In sum, DPL can be characterized as follows: (1) it is a budget support funding instrument with policy reform conditions; (2) the policy reforms focus on sustainable development (poverty reduction and growth instead of macroeconomic stabilization and adjustment); and (3) the policy reforms are identified through a process that is owned by the recipient but coordinated through participation of the development partners.

The following sections show how this modality was applied in Indonesia and Vietnam for climate change related policy reforms and reveal how the aforementioned challenges were addressed.

Case study: climate change program loan in Indonesia

Background: climate change and Indonesia

Vulnerability

According to the official report of the Government of Indonesia (GOI), surface air temperature increased by around 0.5 °C during the twentieth century (BAPPENAS,

2010b).[14] Using Global Circulation Models, the same report estimates that the average temperature in Indonesia will increase by 0.8–1 °C during the period from 2020 to 2050, relative to the baseline period of 1961 to 1990. Directly influenced by the surface air temperature, average sea surface temperature of Indonesian waters is projected to increase as much as 0.65 °C by 2030, 1.10 °C by 2050, 1.70 °C by 2080, and 2.15 °C by 2100. Although the observed and projected rainfall patterns in Indonesia vary widely from region to region and season to season, in general, it is anticipated that rainfall will decrease in the dry season and increase in the wet season (BAPPENAS, 2010b). Sea level has been observed to rise in Indonesia at the rate of 0.2–1 cm per year depending on the location (MOE, 2010) and it is estimated that it will continue to rise at the rate of 0.6 cm per year.[15] Studies indicate that as temperature increases, the frequency of El Niño, which is associated with extreme drought, also increases (MOE, 2010). El Niño had previously been observed once every three to seven years but is already appearing at a rate of once every two to six years (BAPPENAS, 2010b).

Comprising over 17,000 islands and with more than 80,000 km of coastline, Indonesia is highly vulnerable to the above predicted changes to the climate. Rises in sea level rise will directly impact the coastal zones, which are home to 50–60 percent of the population (BAPPENAS, 2010b). Serious consequences can be expected, especially to those whose livelihood depends on fishing. Changes in average temperature and rainfall patterns can also affect crop production, with reductions of 4.6 percent for rice, 20 percent for maize, 65.2 percent for soy, 17.1 percent for sugar, and 21.4 percent for palm oil anticipated by 2050.[16]

Emissions

Taking into account emissions of carbon dioxide caused by loss of forests and devastation of peatlands, in 2005, Indonesia was the world's fourth largest emitter of greenhouse gases after China, the United States, and Brazil (million tons of carbon dioxide equivalent) (World Resources Institute, 2011). According to the emissions summary compiled by UNFCCC and Indonesia Second National Communication on Climate Change to the UNFCCC (SNC),[17] GHG net emissions with land use change and forestry (LUCF) expanded from 498,310 Gg CO_2e in 1994 to 1,377,983 Gg CO_2e in 2000. Emissions further increased to 1,791,372 Gg CO_2e in 2005.[18] The main contributing sectors were LUCF, followed by energy and peat fire (Table 8.1) (MOE, 2010).

The rapid increase of emissions was largely a consequence of increasing energy demands associated with economic growth; coal consumption expanded along with economic growth, which led to increased emissions of GHG. The energy sector emitted 280,938 Gg CO_2e in 2000, which had increased to 369,800 Gg CO_2e by 2005 (MOE, 2010). World Resources Institute (2011) ranked Indonesia the fifteenth largest energy related emitter in 2007.

Recognizing the situation, GOI has studied emission reduction measures and their abatement costs for several sectors, namely peat, forestry, agriculture, power,

TABLE 8.1 Summary of GHG emissions from 2000–2005 for all sectors (Gg CO_2e)

Source	*2000*	*2001*	*2002*	*2003*	*2004*	*2005*
Energy	280,937.58	306,774.25	327,910.62	333,950.21	372,123.28	369,799.88
Industrial process	43,043.52	49,810.15	43,716.26	47,901.63	47,985.20	48,733.38
Agriculture	75,419.73	77,500.80	77,029.94	79,828.80	77,862.54	80,179.31
LUCF	649,254.17	560,546.00	1,287,494.79	345,489.33	617,423.23	674,828.00
Peat fire	172,000.00	194,000.00	678,000.00	246,000.00	440,000.00	451,000.00
Waste	157,327.96	160,817.76	162,800.37	164,073.89	165,798.82	166,831.32
Total with LUCF and peat fire	1,377,982.96	1,349,448.96	2,576,951.98	1,217,243.86	1,721,193.07	1,791,371.89
Total without LUCF and peat fire	556,728.79	594,902.96	611,457.19	625,754.53	663,769.84	665,543.89

Source: MOE (2010).

transportation, cement production, and buildings (DNPI, 2010). The study indicates that, under a business-as-usual scenario, Indonesia's 2005 emissions would grow at a rate of 2 percent annually and reach 2.5 Gt CO_2e in 2020 and 3.2 Gt CO_2e in 2030 (Figure 8.1).

Under these circumstances, countermeasures to reduce emission of greenhouse gases by mitigating the loss of forested areas, or developing renewable energy as well as conserving energy have become an urgent issue for Indonesia.

GOI responses as a basis of CCPL (2007–2009)

Although Indonesia is not an Annex I country of the UNFCCC, its large emissions footprint gives it an important position in worldwide efforts to fight climate change. Moreover, recognizing that Indonesia is one of the most vulnerable countries in the world to climate change, GOI is committed to playing an active role in diverse range of international negotiations to push for collective action. This leadership is exemplified by them hosting the COP13 of the UNFCCC in Bali, which generated the Bali Action Plan.[19]

In addition to playing a key role internationally, the GOI has been strengthening its domestic institutions. Specifically, it developed during 2007 two key documents in time for the COP13. The National Action Plan on Climate Change (NAP-CC), which was published in November 2007, contains initial guidance for a multi-sector coordination effort designed to address jointly the challenges of mitigation and adaptation to climate change. Following this, the National Development Planning Agency (BAPPENAS) of GOI published in December 2007 (revised in July 2008) the "National Development Planning: Indonesian Responses to Climate Change (Yellow Book)," which strengthened and reinforced the National

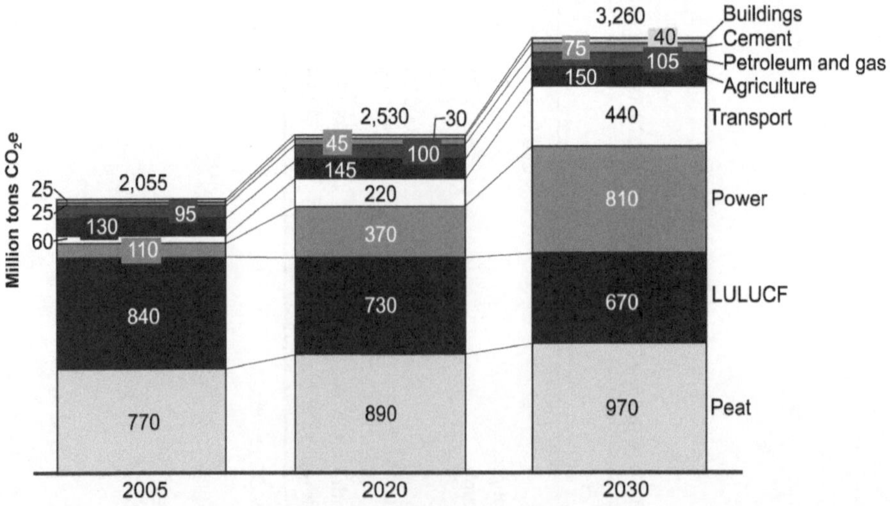

FIGURE 8.1 Indonesian emission growth projection from 2005 to 2030 (source: DNPI, 2010).

Medium-Term Development Plan (RPJMN) 2004–2009 as well as provided input to guide the integration of climate change considerations into the next RPJMN targeting 2010–2014.

Development of CCPL and its objective

Recognizing the GOI's active role internationally and domestically to fight climate change, Japan began preparation of the Climate Change Program Loan (CCPL) in 2007. There were specifically three developments that led to the formulation of CCPL. The first development came in the form of international politics surrounding climate change. In May 2007, then Japanese Prime Minister Shinzo Abe announced the "Cool Earth 50" initiative at a dinner party with Asian leaders (Office of Prime Minister, 2007). The initiative proposed all countries to set long-term targets with the intent of halving global emissions by 2050 from the current level.[20] It also announced that Japan was committed to extending assistance to developing countries that aspired to achieve this goal. The assistance mechanism came into shape when succeeding Prime Minister Yasuo Fukuda announced the "Cool Earth Partnership" at the Davos Conference in January 2008. The "Cool Earth Partnership" committed a total of approximately US$10 billion in financial assistance over a five-year period starting in 2008. This partnership aimed primarily to support developing countries which were making efforts to reduce GHG emissions by utilizing the "Climate Change Japanese ODA Loan" scheme to implement programs and projects.[21]

In August 2007, just a few months after the announcement of "Cool Earth 50," on the occasion of Prime Minister Abe's visit to Indonesia, Indonesian President Susilo Bambang Yudhoyono welcomed the initiative and proposed that it should be one of the key initiatives to be discussed at COP13, which Indonesia would be hosting at the end of that year. They also agreed to deepen their bilateral relationship with a joint program to address climate change, which was announced as the "Indonesia–Japan Joint Statement for Enhanced Cooperation on Climate Change, Environment and Energy Issues"[22] (MOFA, 2007).

The second development was strong fiscal demand from Indonesia. GOI had increased the amount of infrastructure investment to promote economic growth, and had also increased subsidy expenditures to offset inflated oil and food prices. These factors had exacerbated fiscal deficits in Indonesia since 2007 (JBIC, 2007, 2008b). Thus, financial support from development partners was necessary to cover fiscal deficits (Sudo *et al.*, 2009).[23] In fact, Indonesia has been the recipient of a variety of DPLs from the World Bank, Asian Development Bank (ADB), and Japan International Cooperation Agency (JICA) since 2004. These DPLs amounted to US$400 million at the time and increased yearly, reaching US$1.9 billion in 2007,[24] including CCPL.

The third development was the expectation that Indonesia would garner the attention of other development partners based on its active commitment to climate change. By committing to a more advanced activities for dealing with climate change than those adopted by other developing countries, Indonesia expected to eventually receive from development partners the financial and technical supports to achieve the targets stated in NAP-CC and Yellow Book.

Given these three developments, CCPL took shape as a means of supporting implementation of NAP-CC and Yellow Book and thereby strengthening the institutions to deal with climate change. The policy matrix, therefore, was based on these documents to ensure alignment with national development plans. Since NAP-CC's objective was the integration of climate change measures into the national development plan, this concept was also incorporated into CCPL. As a financing mechanism, CCPL promoted not only planning but the integration of climate change measures with financing, or national budget. It was therefore well positioned to serve as a basis for the formulation of the next Mid-term National Development Plan (2010 to 2014).

CCPL formulation process

Although JICA had engaged in various DPLs to Indonesia since 2004, these early DPLs were developed and lead by multilateral banks, World Bank, and ADB. JICA's participation was as a co-financer. CCPL was the first DPL developed on JICA's own initiative and under its leadership. Accordingly, it dispatched multiple contact missions in 2007 and fact-finding missions in 2008 in order to design an effective approach to policy matrix development as well as to consider the implementation and management structure (Figure 8.2).

The process of CCPL can be summarized as follows: (i) the financers (GOJ (Government of Japan), JICA, and French Development Agency (*Agence Française de Développement* or AfD)) and GOI formulate and agree on the policy matrix, which contains policy actions for three years;[25] (ii) GOI implements and monitors

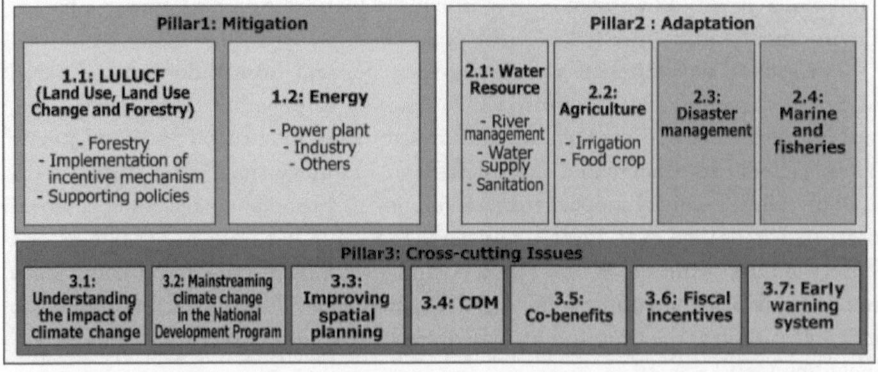

FIGURE 8.2 Structure of CCPL policy matrix (2007–2009) (source: JICA, 2009).

Note
The two sectors (disaster management and marine fisheries) were added in 2009.

the progress of the policy actions with support from the co-financers; (iii) the financers and GOI confirms the implementation on a yearly basis; and (iv) if confirmed, JICA and AfD provide loans that are incorporated into the GOI general budget to cover fiscal deficits. This means that the first tranche of the CCPL is disbursed based on the achievement of prior actions of 2007 and the commitment of policy actions for 2008 and 2009; the second and third tranche are disbursed based on the achievements of policy actions for 2008 and 2009, respectively. After the first three years, the financers consider the further provision of CCPL based on the achievement of the policy actions as well as GOI's requests for funding.

The policy matrix mentioned above is formulated based on discussions with the GOI. As shown in Table 8.2, the matrix comprises three areas, or "pillars," mitigation, adaptation, and cross-cutting issues (JBIC, 2008b; JICA, 2009).

As the result of the above process, during the G8 Summit Outreach Session on July 9, 2008, Japanese Prime Minister Fukuda, expressed his appreciation for Indonesia's contributions to the successful COP13 and pledged the provision of CCPL for the first year.[26] Based on this pledge, JBIC (now JICA) signed a loan agreement for US$300 million with the GOI in September 2008. AfD also provided US$200 million in co-financing in 2008. In 2009, based on satisfactory results and commitment for the following year, JICA and AfD provided a further US$400 million and 300 million, respectively.

Management and implementation arrangement

Because CCPL require strong coordination with all line ministries as well as with co-financers, JICA and GOI held extensive discussions on appropriate management and implementation arrangements. The first issue to be determined was who should be the leading (or executing) ministry. After thorough consideration and discussions, GOI appointed BAPPENAS as the leading ministry. BAPPENAS was the coordinating agency for all line ministries that gives instruction and suggestions on the planning, budgeting, implementation, and evaluation of their programs and projects. There were two specific reasons.

First, BAPPENAS had experience as the leading ministry in the development of both NAP-CC and Yellow Book. Thus, BAPPENAS had demonstrated the ability to coordinate, monitor, and make instructions to other line ministries. Second, since the main purpose of CCPL was to promote the integration of climate change measures into national development plans, BAPPENAS was in the best position to achieve this objective, since it is also the ministry in charge of the national development plans. For these reasons, BAPPENAS was assigned to closely coordinate and monitor the implementation of CCPL policy actions to be implemented by other ministries, including the Ministry of Forestry, Ministry of Mining and Energy, Ministry of Industry, Coordinating Ministry of Economic Affairs, Ministry of Public Works, Ministry of Agriculture, Ministry of Environment, Coordinating Agency for Surveys and Mapping (Bakosurtanal), BMKG, and Ministry of Finance.[27]

TABLE 8.2 Baseline and Main Policy Reform (2007–2009)

Sector	*Baseline (2007)*	*Main Policy Reform (2007–2009)*
Mitigation		
Forestry	• Lack of incentives and management systems for the preservation of forests	• Built the concept of countermeasures in response to REDD,* and formulated detailed implementation mechanism and plans of REDD pilot project
Energy	• Insufficient systems relating to the development of renewable energy	• Institution of energy laws aimed at the promotion of the development of renewable energy, and issuance of incentive measures for renewable energy development, including geothermal
	• Insufficient systems aimed at promoting energy conservation	• Implementation of energy audit and introduction of systems of labeling for energy-saving products • Formulation of sectoral road maps for reduction of carbon dioxide emissions through energy saving • Issuance of ministerial regulation for energy conservation
Adaptation		
Water resources	• Insufficient systems relating to integrated river basin management	• Establishment of government regulations and presidential decrees aimed at integrated watershed management • Formulation of integrated river basin management plans for major river basins
Irrigation	• Deterioration of irrigation facilities etc., due to ineffective management of water for irrigation use and insufficient management for facilities	• Strengthening of systems for the maintenance of management facilities for irrigation water
Agriculture	• Insufficient management of irrigation water and facilities, and insufficient agricultural advisory services	• Establishment and implementation of agricultural advisory programs to respond to climate change
Disaster management**	• Drafting of a national disaster management plan	• Establishment of a national disaster management plan
Marine and fisheries**	• No formulated national plan for the conservation of coral reefs	• Formulation of a national action plan for marine conservation

TABLE 8.2 Continued

Sector	*Baseline (2007)*	*Main Policy Reform (2007–2009)*
Mainstreaming	• Formulation of a national action plan on climate change	• Establishment of a national action plan relating to climate change, and reflection of climate change policies in national development plans (2010–2014)
CDM	• Formulation and approval of CDM projects (13 projects)	• Formulation and approval of clean development mechanisms, and increase in cases of approval of CDM projects (24 projects in 2007 → 70 projects in 2008, in 2009, 90 projects in total)
Early warning system	• Installation of weather observation facilities (automatic weather stations at 47 locations, and weather radars at seven locations)	• Enhancement and expansion of observation facilities aimed at strengthening climate observation structures

Source: JBIC (2008a), JICA (2009, 2010).

Notes

* Reduced Emissions from Deforestation and Degradation.

** Policy actions in the two sectors added in 2009 (disaster management and marine fisheries).

In order to fulfill this role effectively, BAPPENAS established a Steering Committee (SC) for CCPL.[28] The SC convenes twice a year with delegates from GOI, GOJ, JICA, and AfD. SC convenes: (1) to confirm and approve the progress of policy actions; (2) to discuss potential improvements and additional actions to improve the progress; and (3) to approve new policy actions. Echelon 1 Indonesian government officials (vice minister level) from line ministries are expected to attend SC meetings. In particular, Echelon 1 of BAPPENAS, Ministry of Finance, Coordinating Ministry of Social Welfare, and Coordinating Ministry of Economic Affairs rotate as SC chair. Co-financers are also co-chairs, and high-level officers such as an Executive Policy Advisor of the Advisory and Monitoring team (described below), director-level representative of the Japanese Ministry of Foreign Affairs, Minister of Embassy of Japan, Chief Representative of JICA Indonesia, Country Director of JICA head office, attend every SC meeting. Furthermore, a Technical Committee Meeting (TCM)[29] is also held on a regular basis prior to SC meetings. The purpose of the TCM is to discuss the progress of policy actions and other relevant issues as well as to draft new policy actions with technical level personnel (Echelon 2) for reporting of TCM results to the SC (Figure 8.3).

To assist BAPPENAS and the line ministries conduct the day-to-day monitoring as well as implement policy actions, JICA formulated an Advisory and Monitoring team (A&M team), comprising an Executive Policy Advisor (EPA), Task Manager (TM), Sector Experts, and Coordinator.[30,31] The A&M team serves as a

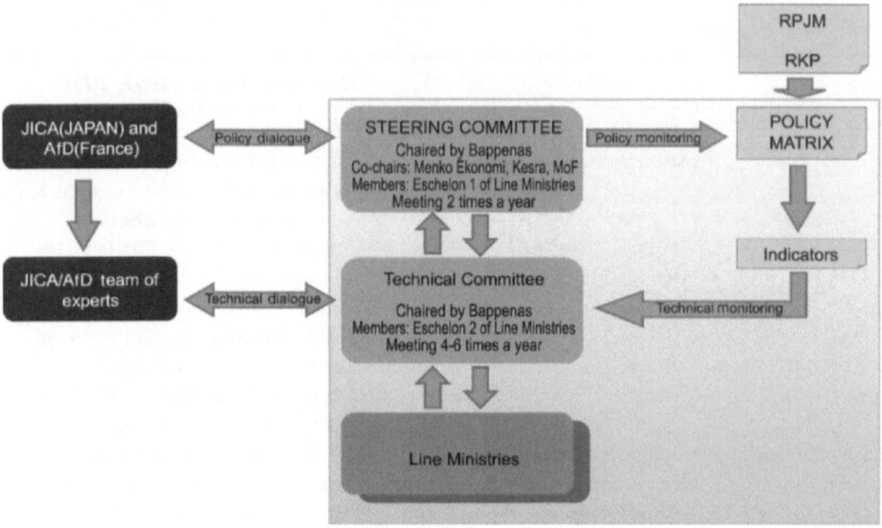

FIGURE 8.3 Implementation structure of CCPL (source: presentation made by BAPPENAS at the first Steering Committee Meeting in 2009).

day-to-day liaison between the line ministries, co-financers (JICA and AfD), and technical experts from JICA and other donors. At appropriate times, such as SC meetings, the A&M team reports on achievement and progress of policy actions, and makes suggestions on necessary actions for improvement.

Previously, we have seen that ownership is a critical element for successful DPL. It is, then, appropriate to look at the various incentives of the participating ministries because without these incentives, be they economic or political, it is difficult to ascertain ownership. The incentive for the Ministry of Finance is obvious. Being the direct beneficiary, it can use funding provided through CCPL to cover fiscal deficits. Although not a direct beneficiary in terms of funding, the incentive of BAPPENAS is also clear, since CCPL facilitates and complements the implementation of national climate change policies, which is the BAPPENAS mandate. The incentives of the line ministries are less direct. By participating and committing to policy reforms under CCPL, line ministries can appeal to the Ministry of Finance for additional budget to implement those policy actions and related projects. In cases where discussions with development partners result in the identification of a particular need, line ministries can expect financial assistances from the development partner for that project. Thus, the ministries and agencies involved in this process exhibit a diverse range of motivation.

Results of CCPL Phase 1 (2007–2009) and toward CCPL Phase 2 (from 2010)

With implementation of the 1st Phase of CCPL (2007–2009), GOI has made considerable progress toward strengthening its institutional frameworks and mobilizing intellectual, financial, and human resources from a variety of government and non-government agencies. In particular, there are seven key developments worthy of enumeration. First, the National Council for Climate Change (NCCC), an inter-agency council for coordinating climate policies, was established in October 2008[32] with a specific mandate to coordinate diverse climate change related activities into a coherent whole. Second, the Indonesian Climate Change Trust Fund (ICCTF) was established in September 2009 to finance projects and activities in accordance with GOI policies.[33] Third, the Indonesia Climate Change Sectoral Roadmap (ICCSR) was issued in December 2009, which delineates the mainstreaming of a variety of sectors into the national development plan over the next 20 years (2010–2029).[34] Fourth, Indonesia pledged in January 2010 to associate with the Copenhagen Accord and submitted a list of voluntary emission reduction measures, which call for a 26 percent cut in emissions from business as usual (BAU) levels by 2020 or a 41 percent cut with international financial support.[35] Fifth, a National Action Plan to reduce GHG emission – essentially the Indonesian NAMAs – is being prepared (as of May 2011). The plan comprises initiatives to reduce GHG emissions during the period 2010 to 2020 in five major sectors: forestry (including peatland), agriculture, waste, industry, and energy (including transportation). Sixth, the Second National Communication (SNC), which had been drafting since 2009, was submitted to the UNFCCC in November 2010. The SNC provides the up-to-date data on the national situation. Last, GOI initiated actions related to REDD-plus for reducing emission from deforestation and forest degradation such as the UN-REDD National Programme that began in March 2010.

These new developments required additional support, both technical and financial, to implement. GOI and co-financers, therefore, began discussing the second phase of CCPL in 2009. The policy matrix for CCPL phase two will target actions from 2010 and will be aligned to ICCSR as well as the draft of the National Action Plan to reduce GHG emissions.

In addition to discussions on the new policy matrix, there was also extensive dialogue on improving the CCPL process based on lessons learned during phase one. Of these, one key lesson was that the length of phase one policy matrix made it burdensome to monitor and to ensure quality of each policy action. JICA and the co-financers agreed with GOI to reduce the number of line items in 2010 policy matrix, thereby enabling both cost-efficient monitoring and better ownership of actions by Indonesian officials involved. This streamlined approach also placed more emphasis on priority climate change policies.

Agreement on the new policy matrix was reached between GOI and the co-financers in April 2010. JICA provided US$300 million in June 2010, based on the achievement during 2009 and commitment for 2010 actions.[36] In addition, AfD provided US$300 million and newly joined World Bank provided US$200 million.

The new policy matrix also comprises three pillars: key policy issues for climate change, mitigation, and adaptation (JICA, 2010). Compared with the phase one policy matrix, the new policy actions focused on GOI upstream policies, mitigation

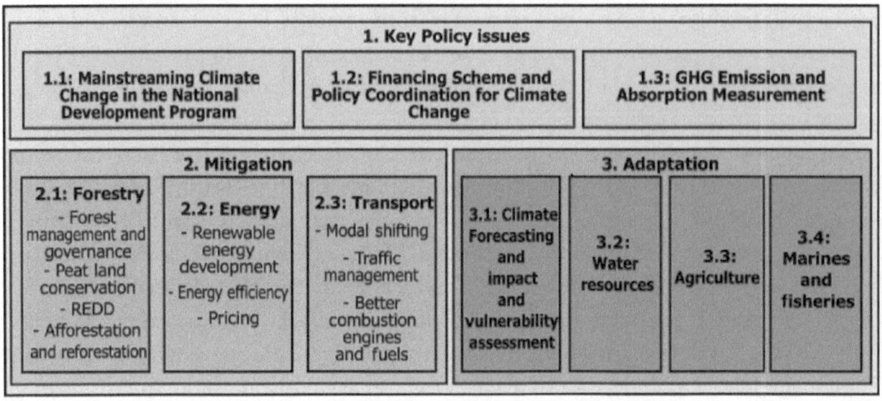

FIGURE 8.4 Structure of CCPL policy matrix (2010–) (source: JICA, 2010).

TABLE 8.3 Main policy actions (2010)

Sector	*Main policy actions (2010)*
Key policy issues for climate change (upstream strategies)	
Mainstreaming	• Issuance of the national action plan to reduce emissions by 26% by 2020 as a presidential regulation • Establishment of funding mechanisms for the implementation of climate change policies • Submission of Second National Communications to UNFCCC
Mitigation of global warming by greenhouse gas absorption and emissions control	
Forestry	• Establishment of management standards and procedures for forest management units (FMU) • Strengthening of cross-ministerial coordination for the reduction of GHG emissions from peatland
Energy	• Expansion of incentive mechanisms for the promotion of renewable energy development • Formulation of energy conservation master plan
Transportation	• Acceleration of modal shift in transportation by the expansion of bus rapid transport system
Strengthening of adaptation capacity to deal with the adverse effects of climate change	
Climate forecasting and impact and vulnerability assessment	• Development of climate change modeling as the basis for impact and vulnerability assessment
Water resources	• Continued formulation of integrated river basin management plans for major river basins
Agriculture	• Continued implementation of agricultural advisory programs that deal with climate change and water conservation, and plans for efficient use of farm lands

Source: JICA (2010).

policies that contribute to achieving the 26 percent target for reduction of GHG emissions by 2020, and the MRV mechanism, as promoted by the Hatoyama Initiative (MOFA, 2009).

Case study: support program to respond to climate change in Vietnam

Background: Vietnam and climate change

Vulnerability

According to the official climate change scenario of the Government of Vietnam (GOV), the average temperature in Vietnam rose by 0.5–0.7 °C between 1958 and 2007.[37] Using B1, B2, and A2 emission scenarios of the IPCC to estimate the low-, medium-, and high-emission scenarios for Vietnam, it is estimated that by 2100 the average temperature will rise 1.1–1.9 °C (low), 1.6–2.8 °C (medium), and 2.1–3.6 °C (high) when compared to 1980–1999 level. Although there are no statistically significant trends from the past, the same report also estimates that by 2100 precipitation will decrease by 2.9–9.4 percent (low), 4.4–17.4 percent (medium), and 5.6–22 percent (high) during the dry months of March to May compared to 1980–1999 level. For the same years, the precipitation will increase by 0.2–9.5 percent (low), 0.3–15.1 percent (medium), and 0.3–19.1 percent (high) during the wet months of June to August. Based on data from tidal gauges along the coast, it was observed that sea level rose by 3 mm per year, between 1993 and 2008. It is also estimated that a 65–100 cm rise in sea level by 2100 would inundate 13–38 percent of the Mekong Delta. The report also reveals that typhoons are increasing both in frequency and intensity (MONRE, 2009). Although no evidence is presented in the report, this is probably a reasonable assumption, given that typhoon activity increases as sea surface temperature rises (World Bank, 2011a).[38]

Many characteristics of Vietnam make it highly vulnerable to the above climate change phenomena, especially sea level rise. With a coastline over 3,200 km long and 50 percent population rate along low-lying coastal zones,[39] it is estimated that 10.8 percent of the population would be displaced and 10 percent of the GDP lost should sea level rise by 1 meter (Dasgupta *et al.*, 2007). High reliance on agriculture is another characteristic that leaves Vietnam vulnerable to climate change. Rural population, who are dependent mainly on agricultural activities, account for 80 percent of the entire population and agricultural output contributes to 20 percent of the GDP (World Bank, 2011b).[40] Given these facts, Vietnam is listed as the thirteenth most vulnerable of the 170 countries assessed in the Climate Change Vulnerability Index published by a global risk advisory firm, Maplecroft. The Mekong Delta, in particular, is identified as one of three global hotspots in terms of potential population displaced as a result of sea level rise (IPCC, 2007).

Emissions

Vietnam's economy grew rapidly throughout the 1990s and 2000s. Average annual GDP growth was 7.9 percent during the 1990s and continued to grow 7.6 percent annually from 2000 to 2009 (World Bank, 2011b). Accordingly, energy production and consumption showed annual increases of 14.8 percent and 11.6 percent, respectively, during the period from 1990 to 2005 (JICA, 2008). As of 2005, there were 19 million registered motorbikes, which is almost ten times more than in 1990 (JICA, 2008).

Given this rapid economic growth, GHG emissions also surged. According to official national inventories in 1994 and 2000, GHG emissions, including offsets, increased by 45.3 percent during the period from 1994 to 2000 (Table 8.4).[41] According to the World Resources Institute, Vietnam's emissions reached 177 million tons of CO_2e in 2005 (excluding offsets), which makes it the world's thirty-fifth largest emitter contributing 0.5 percent of the global emissions (World Bank, 2011a). Although Vietnam is not currently a large contributor in absolute terms, its emissions growth rate is one of the fastest in Asia. It is estimated that by 2030 Vietnam's emissions will reach nearly 600 million tons of CO_2e, giving it a footprint roughly the same size as current emissions from Indonesia (World Bank, 2011a).

GOV responses

Recognizing the seriousness of potential impact of climate change to Vietnam, the GOV ratified the UNFCCC on November 16, 1994 and the Kyoto Protocol (KP) on September 25, 2002.[42] Foreseeing potential economic benefit immediately after ratification of the KP, the GOV quickly began laying the foundation for domestic institutions needed to utilize the mechanisms available under the KP. The Designated National Authority for CDM was assigned in March 2003[43] and the National Executive Board and Consultative Board for CDM were established in April of the same year.[44] Since then, the GOV has drawn up a variety of action plans and guidelines as well as institutionalized financial incentives for promoting investments in CDM projects.[45] It was, nevertheless, not until 2009 that these efforts blossomed. As late as the end of 2008, there were only two registered CDM projects: the Rang Dong Oil Field Associated Gas Recovery and Utilization Project and the Song Muc Hydro Power Station Regeneration Project. Since that time, 54 other projects have been registered at the CDM Executive Board.

In addition to the CDM, the GOV has initiated numerous strategies and activities that have climate change as an objective. These include the National Strategy for Natural Disaster Prevention, Response and Mitigation to 2020,[46] the Forestry Development Strategy 2006–2020,[47] the National Energy Efficiency Program 2006–2015,[48] and the National Scheme on the Development of Biofuels up to 2015, with a vision to 2020.[49]

A move toward comprehensive action to fight climate change began while the Conference of the Parties to the UNFCCC was meeting in Bali. During a regularly scheduled government meeting, the Prime Minister instructed the Ministry of Natural Resources and Environment (MONRE) to develop a national target program to respond to climate change.[50] In Vietnam, national target programs are inter-ministerial and budgeted programs. After a year of background study and domestic and international consultation, the GOV issued the National Target Program to Respond to Climate Change (NTP-RCC) in December 2008.[51]

The main objectives of the NTP-RCC are to assess the impact of climate change on a variety of sectors and to develop feasible action plans in response to the impacts. It identifies seven tasks, namely: (1) assessment of the impact of climate change and development of climate change and sea level rise scenarios; (2) development of science and technology programs on climate change; (3) capacity building; (4) raising public awareness and training human resources; (5) enhancement of international cooperation; (6) development of a standard framework for mainstreaming climate change into development programs; and (7) development of action plans at the ministerial, sectoral, and provincial levels to respond to climate change. Within these tasks, the NTP-RCC lists 26 activities to be completed by 2015. The total cost for these activities were estimated to be VN$1,965 billion or about US$110 million.

Moreover, the Prime Minister has instructed MONRE to lead the development of a climate change national strategy and NAMA[52] while implementing NTP-RCC and elaborating the action plans. In addition, the Vietnamese Ministry of Planning and Investment (MPI) is currently leading the development of a low carbon growth strategy with assistance from the UK Department for International Development (DFID) and the World Bank.[53]

Development of SP-RCC and its objective

The NTP-RCC provided a foundation on which the international development community could stand in joining the GOV effort to fight climate change. It showed GOV's commitment and ownership which are the most critical factors in successful assistance, as mentioned above. Within one month of the issuance of the NTP-RCC, Denmark became the first country to commit, granting approximately US$40 million to finance its implementation in the form of sector budget support.[54] Two months later, in March 2009, JICA and the AfD expressed their intent to support to the NTP-RCC[55] and selected DPL as the modality based on the recognition that institution building should take priority over concrete investments. This support program was named the Support Program to Respond to Climate Change (SP-RCC).[56]

SP-RCC had two broad objectives. First, it aimed to build the institutions – policies, regulations, etc. – necessary to promote effective and efficient climate change responses including both mitigation and adaptation measures. Second, by opening up the policy dialogue to all concerned development partners and NGOs, it served as a platform for donor coordination and aid harmonization. Given the inherent complexity of climate change, a platform for donor coordination was

TABLE 8.4 GHG emissions of Vietnam, 1994 and 2000

Sector	*GHG emissions (mil. tons CO2e)*				*Change 1994 to 2000*		
	1994		*2000*		*Volume*	*% total*	*% Δ*
	Volume	*% total*	*Volume*	*% total*			
I Energy	*25.64*	*24.7*	*52.77*	*35*	*27.1*	*57.7*	*105.8*
1 Fuel combustion	24.66	23.7	45.90	30.4	21.2	45.1	86.2
A Energy industry	4.13	4.0	11.17	7.4	7.0	15.0	170.5
B Manufacturing/construction	7.71	7.4	15.02	10.0	7.3	15.5	94.9
C Transportation	3.66	3.5	11.89	7.9	8.2	17.5	225.0
D Commercial	1.99	1.9	2.96	2.0	1.0	2.1	48.9
E Residential	4.78	4.6	2.31	1.5	–2.5	–5.2	–51.6
F Agriculture/forestry/aquaculture	0.89	0.9	1.37	0.9	0.5	1.0	53.9
G Others	1.50	1.4	1.17	0.8	–0.3	–0.7	–21.8
2 Fugitive	0.98	0.9	5.04	3.3	4.1	8.6	413.7
A Solid fuel	0.83	0.8	1.87	1.2	1.0	2.2	124.6
B Gasoline and oil	0.15	0.1	3.17	2.1	3.0	6.4	2051.8
II Industrial processes	*3.81*	*3.7*	*10.01*	*6.6*	*6.2*	*13.2*	*162.8*
1 Cement	–	–	6.63	4.4	–	–	–
2 Lime	–	–	0.82	0.5	–	–	–
3 Ammonia	–	–	0.01	0.0	–	–	–
4 Carbide	–	–	0.01	0.0	–	–	–
5 Steel	–	–	2.54	1.7	–	–	–

III Agriculture	*52.45*	*50.5*	*65.09*	*43.1*	*12.6*	*26.9*	*24.1*
1 Enteric fermentation	7.07	6.8	7.73	5.1	0.7	1.4	9.3
2 Manure management	2.71	2.6	3.45	2.3	0.7	1.6	27.2
3 Rice cultivation	32.75	31.5	37.43	24.8	4.7	9.9	14.3
4 Agricultural soils	8.06	7.8	14.22	9.4	6.2	13.1	76.4
5 Burning of savannas	0.40	0.4	0.59	0.4	0.2	0.4	47.7
6 Burning of agricultural residues	1.46	1.4	1.67	1.1	0.2	0.5	14.6
IV LULUCF	*19.38*	*18.7*	*15.10*	*10*	*–4.3*	*–9.1*	*–22.1*
1 Changes in biomass stocks	–39.27	–37.8	–49.83	–33.0	–10.6	–22.4	26.9
2 Forest and grassland conversion	60.88	58.6	43.91	29.1	–17.0	–36.1	–27.9
3 Abandonment of managed lands	–11.05	–10.6	–7.33	–4.9	3.7	7.9	–33.7
4 CO_2 uptake/emission from soils	8.82	8.5	28.36	18.8	19.5	41.5	221.5
V Waste	*2.57*	*2.4*	*7.93*	*5.3*	*5.4*	*11.4*	*209.0*
1 Solid waste	1.39	1.3	5.60	3.7	4.2	8.9	302.0
2 Wastewater	0.02	0.0	0.03	0.0	0.0	0.0	31.7
3 Industrial wastewater	0.02	0.0	1.34	0.9	1.3	2.8	7953.0
4 Human waste	1.13	1.1	0.96	0.6	–0.2	–0.4	–15.0
Total emissions	*103.84*	*100*	*150.90*	*100*	*47.06*	*100*	*45.3*

Sources: MONRE (2003, 2010).

especially necessary. Unlike typical sectors such as transportation or power, climate change (1) is not a separate agenda but rather is embedded into all other sectors, (2) attracts a large number of development partners, both multilateral and bilateral, government and non-government, or financial and technical, and (3) is financed by numerous (and many more to come) international financing facilities each with its own procedures, objectives, and scopes.

SP-RCC process

As previously mentioned, one of the crucial elements of a DPL is the process by which policy reforms are selected and agreed to. This process must ensure ownership of the recipient country, without which reforms are not implemented, as we learned from the failure of the SAL. There is, however, no standard DPL process for ensuring ownership and each DPL must be tailored to its own specific context. For SP-RCC, this specific context was twofold. First, the leading agency was the ministry in charge of environmental issues: MONRE. Similar to environment ministries in other countries, MONRE's political strength was relatively limited. Partly due to this and partly due to the broadness of its mandate, MONRE was short on human resources, both in quantity and quality. The procedure, therefore, must ensure ownership but should also take into consideration the political and administrative limitations of MONRE. Second, the program that SP-RCC was meant to support, the NTP-RCC, was not just a political document. It had its own management and implementation mechanisms. Thus, there was great risk of creating a duplicate mechanism unless due care was taken. Given this situation, the procedure developed as follows.[57]

At the beginning of the year, development partner representatives in Vietnam including iNGOs were invited to attend a two-week-long series of donor coordination meetings. Separate donor coordination meetings were held for each climate change topic, such as energy and climate change or water and climate change. Each meeting lasted two to three hours, with a typical agenda including (1) a presentation by the lead donor describing their assessment of the sector and the priority policy agenda, (2) comments from other development partners describing their current activities and their perception of the priority policy agenda,[58] and (3) discussion and conclusion.[59] After the two weeks, the donors' proposals for key policy reforms are consolidated.

Shortly after the donor coordination meetings, meetings with GOV ministries – called technical meetings – are organized. Similar to the donor coordination meetings, separate meetings lasting two to three hours are held for all sectors and are attended by GOV representatives from related departments at different ministries. For example, participants to the water sector meeting will be the Water Resource Department from MONRE, Water Management Department from MARD (Ministry of Agriculture and Rural Development), Technical Infrastructure Department from MOC (Ministry of Construction) and the DHMCC (Department of Hydrometeorology and Climate Change) of MONRE. A tentative policy matrix is sent to the ministries prior to the meeting; therefore, the agenda for a typical technical meeting would start with the comments from the ministries regarding the policy matrix, followed by technical discussions. At the end of two weeks, a list of agreed-to policy reforms, or the policy matrix, is compiled and submitted for authorization to higher authorities in the GOV. During this initial process, requirements for technical assistance are also identified[60] together with policy reforms.

Given this commitment, another series of meetings is held in the middle of the year to monitor progress and another is held at year-end to review the achievements of the policy reforms. Based on the results of this year-end review, financing is provided by the development partners in the form of general budget support. A review of 2009 achievements resulted in the provision of JPY10 billion (around US$120 million) and €20 million soft loans from JICA and AfD, respectively. A review of 2010 achievements was undertaken in January 2011, and financial arrangements from JICA, AfD, World Bank and CIDA (Canadian International Development Association) are currently under preparation.

Management and implementation arrangement

Insofar as possible, the process described above is managed and implemented using existing NTRP-RCC mechanisms. NTP-RCC is supervised by the National Steering Committee (SC) and the Executive Board (EB). The SC comprises ministers from the relevant ministries and is headed by the prime minister; EB comprises vice ministers from relevant ministries and is headed by the Minister of Natural Resources and Environment. Day-to-day operations are implemented by the secretariat, NTP-RCC Standing Office (SO), which is located inside MONRE. SP-RCC is integrated into this structure by adding SP-RCC management responsibilities to SC and EB mandates. Day-to-day operations of the SP-RCC are implemented by a newly created Program Coordination Unit (PCU), which comprises some overlapping members of the SO. Overlapping members ensure the integrity of SP-RCC and NTP-RCC at the operational level. Recognizing the limited capacity of MONRE, technical assistance is provided by JICA to the PCU in order that this program be managed effectively.

Financial mechanism

One notable aspect of SP-RCC is a supplementary mechanism for setting out the rules by which ministries are able to access the funding provided through SP-RCC. SP-RCC is a general budget support; therefore, there are no requirements from the financers regarding the use of funding.[61] In other words, once the funding is entered into the general budget, the GOV can spend it freely in accordance with its own regulations; the financers do not track how the funds are actually spent. Nevertheless, the GOV chose to create this supplementary mechanism in order to accomplish two objectives. First, to link the use of the funds provided through the program to

TABLE 8.5 Main policy reforms for the first and second year

Mitigation

Energy

- Adopt the master plan for renewable energy development and issue regulation to encourage renewable energy development.
- Prepare regulations to enforce the Law on Energy Efficiency and Conservation

Transportation and construction

- Adopt the master plan for promoting public bus system including promoting CNG/LPG bus operation
- Revise energy efficiency building code

Forests, land use, and agriculture

- Implement Payment for Forest Environmental Services (PFES) pilot policy in several provinces

CDM

- Strengthen legal framework for promoting investments to CDM projects

Adaptation

Water

- Develop a comprehensive program for strengthening the legal and institutional framework for the governance of water resources

Disaster risk management

- Monitor and support implementation of the provincial disaster prevention plan

Forests, biodiversity, and agriculture

- Promote research and assessment for climate change's impacts on crop production in order to form basis for developing active measures against climate change
- Improve the management of special use forest
- Develop implementing regulation of Law on Biodiversity

Health

- Identify areas vulnerable to health impact caused by climate change and create regional maps with consideration of type of potential medical risk

Cross-cutting issues

Analysis and assessment of the state of climate change

- Update climate change scenario analysis report using improved methodologies

Financial mechanisms

- Introduce annual reports on budget allocation for NTP-RCC and budget expenditure estimate with regard to climate change

Mainstreaming climate change in development plans

- Develop the annual development plan for 2011 in which climate change issues are integrated and mainstreamed

the program's primary goal of climate change response. Second, to provide incentives for the participating ministries. In a DPL, or more generally in general budget support, only the finance ministry is able to access directly the allocated financing, which leaves other ministries with little incentive to participate or commit to policy reforms. By creating a mechanism in which financing, either partially or wholly, is channeled to other ministries, they have sufficient incentives to participate.

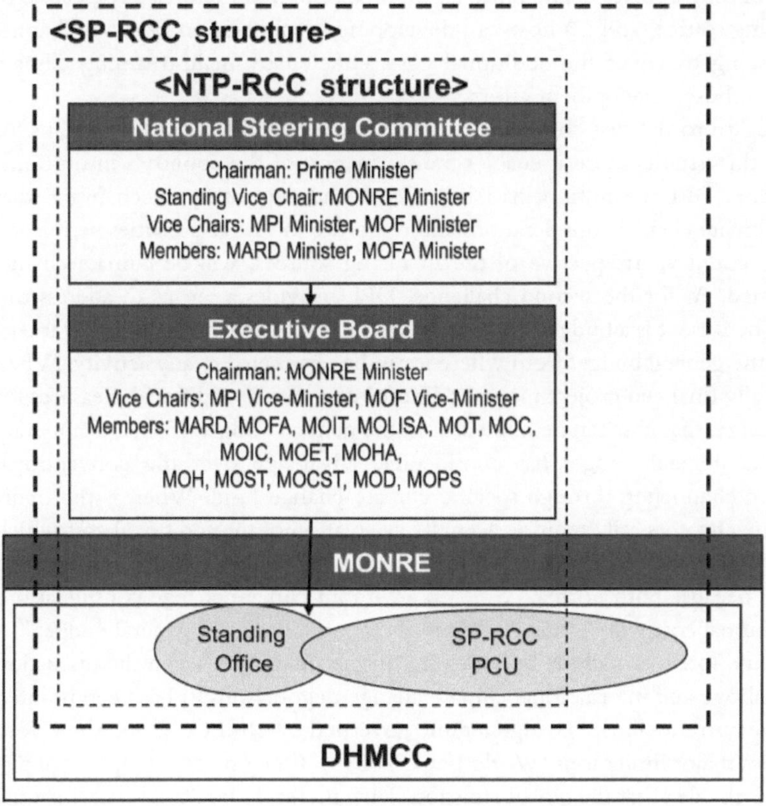

FIGURE 8.5 Management and implementation arrangement of SP-RCC (source: presentation made by JICA and AfD on November 17, 2010 at the climate change donor coordination meeting organized by the UN and Danish Embassy).

On December 10, 2010, after over a year of intra-government discussion, the framework of this financial mechanism was approved by the Prime Minister.[62] This decision allows funding received by the GOV through SP-RCC to be utilized as follows, in order of priority: (1) for the implementation of NTP-RCC; (2) for achieving the agreed-to policy reforms; and (3) for climate change response projects selected according to criteria to be developed by MONRE. The remaining amount will be merged into the general budget.

Conclusion

This review of CCPL in Indonesia and SP-RCC in Vietnam, suggests that DPL is an effective mechanism for addressing the two challenges posed in the introduction, namely: (1) how can development cooperation agencies assist developing countries in mainstreaming climate change into all their development efforts, irrespective of the financing source, and (2) how can development cooperation agencies channel their financing to cover the additional costs entailed by mainstreaming climate change into those development efforts?

With regard to the first challenge, we have seen that DPL provides a means for addressing this issue because it enables transformation of the country's institutional environment. DPL institutionalizes both legal and regulatory mechanisms that promote climate change considerations and responses so that all activities performed within the country, irrespective of the financing sources, will be climate change mainstreamed. As for the second challenge, DPL provides a means to address this issue, too, because it is a budget support. Financing provided through DPL is integrated into the general budget from where it can be channeled to any activity. When a domestically financed project incurs additional cost due to additional measures for climate change, that cost can be funded from the general budget. In fact, channeling through the general budget has considerable advantage over the conventional approach of channeling through specific climate change fund. Whereas the traditional approach necessarily requires accurate calculation of the additional cost which is by no means trivial, channeling through the general budget will not require such calculation because both normal cost – cost without climate change considerations – and additional cost will be financed from the same source, the general budget.[63]

We would like to conclude by touching on the linkage between the discussion presented above and international climate change negotiations. It has been pointed out that the current climate change regime governed by UNFCCC and the Kyoto Protocol has major limitations (World Bank, 2009b). One obvious symptom of this limitation is the fact that the global emissions have increased by 25 percent since the signing of the Kyoto Protocol (World Bank, 2009b). One critical issue toward strengthening of the regime is the treatment of developing countries, since at present only developed countries have legally binding emission targets. It seems unlikely, however, that developing countries will accept emission targets because they are perceived to be limits on development.

One prominent proposal to bridge this divide between developed and developing countries is the multitrack approach (World Bank, 2009b), in which developed countries remain on the "target track" with mandatory emission targets, while developing countries are on the "policy-based track" with only mandatory introduction of climate change related policies. It is said that the policy-based track would be easier to accept for developing countries, because climate change policies can be designed to meet other development objectives such as energy demand management or water use efficiency. That climate change DPLs have been realized in Indonesia and Vietnam is a good indicator that this multitrack approach may well be feasible, because a country like Vietnam, that is vocally reluctant to commit to emission targets, is, on the other hand, fully committed to implementing climate change related policy reforms.

Furthermore, it is easy to see the similarity between climate change DPL and the proposed policy-based track. They both address climate change through policy reform. Whether DPL can be enhanced and linked to the international regime requires further consideration but at the very least experience gained through DPL should be used to design the policy-based track. Specifically, the importance of encouraging the developing country's ownership of the policy reforms cannot be over-stressed.

Notes

1 The views expressed herein are those of the authors and should not be attributed to their affiliated institutions.
2 Senior Expert, International Development Associates.
3 Senior Project Formulation Advisor, JICA Vietnam Office.
4 In the Cancun Agreements, the parties agreed to review and strengthen the long-term goal of below 2 °C to consider 1.5 °C.
5 Vulnerability is "the degree to which a system is susceptible to, and unable to cope with, adverse effects of climate change" (IPCC, 2007). Vulnerability, thus, can be reduced through the reduction of adverse effects (mitigation) and through decreasing of susceptibility (adaptation).
6 Sustainable development can be defined as "a process of change in which the exploitation of resources, the direction of investments, the orientation of technological development and institutional change are all in harmony and enhance both current and future potential to meet human needs and aspirations" (IPCC, 2007). Sustainable development is inclusive of the concept of "low carbon development"; Cancun Agreements stipulate that "social and economic development and poverty eradication are the first and overriding priorities of developing countries and that a low-carbon development strategy is indispensable to sustainable development."
7 "Mainstreaming" is defined as "the integration of policies and measures to address climate change in ongoing sectoral and development planning and decision-making, aimed at ensuring the sustainability of investments and at reducing the sensitivity of development activities to current and future climatic conditions" (IPCC, 2007). At the project level, therefore, it is the integration of measures to address climate change (both mitigation and adaptation) at the various stages of the project cycle: planning, design, construction, and operation. A similar term is "climate proofing"; however, its main focus is on adaptation (ADB, 2005).
8 The term "development policy lending" usually refers to the World Bank's current instrument for policy-based lending. In this chapter, however, the term is used broadly to include reference to the current policy-based lending instruments of other development cooperation agencies such as ADB and JICA. This is because (1) even though different agencies use different terms – DPL for World Bank, "program loan" for JICA, and "program lending" for ADB – they are all similar in characteristics, and (2) the term DPL explicitly distinguishes itself from the previous instance of policy-based lending, SAL.
9 See Chapter 2 of Mosley *et al.* (1991) for a detailed review of the pathway to the development of SAL.
10 This summary of the criticisms is based on (Summers and Pritchett, 1993). Interestingly, one of the most vocal critics of SAL was Japan, the biggest co-financer. See OECF (1991). Based on the lessons from the "miracles" of the East Asian countries, OECF highlighted the importance of the role of the government in leading the country's path to development.
11 See, for example, *Assessing Aid* (World Bank, 1998), *Adjustment Lending Retrospective* (World Bank, 2001) and Dollar and Svensson (2000).
12 World Bank Operational Manual, Operational Policy 8.6, "Development Policy Lending" (August 2004).

13 World Bank Operation Directive 8.6, "Adjustment Lending" (December 1992).

14 Unless otherwise indicated, climate data and descriptions in this section refer to BAPPENAS (2010b).

15 See BAPPENAS (2010a) for a detailed assessment and analysis on air temperature increase.

16 See BAPPENAS (2010b), MOE (2007), MOE (2010) and World Bank (2009a) for more comprehensive assessment of climate change impacts to various sectors in Indonesia.

17 This Indonesian SNC consists of the following information: (i) national circumstances, other information, and constraint/gaps related financial, technical, and capacity needs; (ii) greenhouse gases inventory based on the IPCC 2006 Greenhouse Gas Calculation; (iii) vulnerability assessment and adaptation that focuses on developing climate change adaptation programs; and (iv) greenhouse gases abatement that focuses on developing climate change mitigation programs. See MOE (2010) for detailed information.

18 GHG emissions data for 1994 were extracted from UNFCCC website (http://unfccc.int/files/ghg_data/ghg_data_unfccc/ghg_profiles/application/pdf/idn_ghg_profile.pdf), and 2000 and 2005 data from MOE (2010).

19 This plan articulated the necessary steps to developing a new international climate regime beyond 2012. See http://unfccc.int/files/meetings/cop_13/application/pdf/cp_bali_action.pdf.

20 "Cool Earth 50" proposed the following: (1) setting a long-term target of cutting global emissions by half from the current level by 2050 as a common goal for the entire world and sharing a long-term vision of developing innovative technologies and building a low carbon society, centering on those technologies; (2) "three principles" in designing a concrete framework for addressing global warming beyond 2013, which aims to have the participation of all people on Earth, such as "all major emitters must participate, thus moving beyond the Kyoto Protocol, leading to global reduction of emissions," "the framework must be flexible and diverse, taking into consideration the circumstances of each country," and "the framework must achieve compatibility between environmental protection and economic growth by utilizing energy conservation and other technologies; and (3) a national campaign for achieving the Kyoto Protocol Target. See www.kantei.go.jp/foreign/abespeech/2007/05/24speech_e.html.

21 This partnership aims to support developing countries which are making efforts to reduce GHG emissions and achieve economic growth in a compatible way, on the basis of policy consultations between Japan and those countries. According to the announcement from the GOJ, this finance is provided through several channels, one of the channels to be "Climate Change Japanese ODA Loan" with preferential interest rate. See www.mofa.go.jp/policy/economy/wef/2008/address-s.html.

22 It was announced at top-level conference between Indonesia and Japan in August 2007. See www.mofa.go.jp/region/asia-paci/pmv0708/joint.html.

23 As the Indonesian economy was inflationary, from the standpoint of maintaining the economic balance it was difficult to raise taxes or cut subsidies to ensure financial resources. In contrast, considering the cost of capital procured from the market, receiving financial assistance from donor countries is better for covering fiscal deficits than the issuance of national bonds. See Sudo *et al.* (2009).

24 In 2009, the total amount was over 2.9 billion. The data are based on the information from the Ministry of Finance at the steering committee held in November 2010.

25 In parallel, in order to assist GOI in realizing the policy actions, technical assistance needs are also identified.

26 This became the first instance of Japanese Climate Change ODA under the "Cool Earth Partnership" (MOFA, 2008).

27 Of course, as the borrower of DPL is the Ministry of Finance, the request of CCPL with the proposed loan amount comes from the Ministry of Finance. The function of ministries in Indonesia will be described below.

28 BAPPENAS Ministerial Decree No. 203/2008.

29 Ibid.

30 EPA's responsibilities are: (1) providing policy recommendation to GOI related to the policy matrix; (2) providing policy recommendation to GOJ and JICA regarding Japan's assistances toward climate change in Indonesia; and (3) leading the monitoring and evaluation activities of CCPL. TM and the other team members are responsible for conducting (1) process monitoring and evaluation and (2) technical assistance for specific policy actions.

31 AfD also provided a forestry expert who has been part of the A&M team.

32 Presidential Regulation No. 46/2008. See http://adaptasi.dnpi.go.id/index.php/main/contents/54 for detailed information.

33 In order to gather additional financial support for efforts in climate change, ICCTF gives opportunity to wider development partners to contribute their support to Indonesia climate change program. See www.icctf.org/site/.

34 To elaborate further on the documents mentioned above and also to speed up the implementation by various relevant sectors, BAPPENAS initiated the development of ICCSR. Priority programs of the Roadmap are outlined in four phases of five years each. The purpose of this study is to mainstream climate change into the sectors by integrating it into development planning and in the short-time period is how to mainstream climate change within the national priority and action plan 2010–2014. In addition, the Roadmap will be considered formulating the strategic plans of each ministry and agency. See BAPPENAS (2010b).

35 In a speech to G20 leaders on September 25, 2009, the Indonesian President said the government was crafting a policy that would cut emissions by 26 percent by 2020 from BAU levels, and 41 percent with international financial support. To formalize this commitment, Indonesia has associated with Copenhagen Accord in January 2010, which was the initiation of voluntary emission reduction resulting from the last Climate Change Conference of Parties in Copenhagen. A letter dated January 19, 2010 from Executive Chair National Council on Climate Change of Indonesia to the Executive Secretary of the UNFCCC stated that Indonesia was now in the process of finalizing the Indonesia National Action Plan to achieve 26 percent to 41 percent CO_2e emission reductions. This position was reaffirmed in a letter from Executive Chair National Council on Climate Change of Indonesia to the Executive Secretary of the UNFCCC dated January 30, 2010 that stated the Indonesian voluntary mitigation of 26 percent from its BAU by 2020 through the action of: (i) sustainable peat land management; (ii) reduction in rate of deforestation and land degradation; (iii) development of carbon sequestration project in forestry and agriculture; (iv) promotion of energy efficiency; (v) development of alternative and renewable energy sources; and (vi) shifting to low emission transport. See http://unfccc.int/files/meetings/cop_15/copenhagen_accord/application/pdf/indonesiacphaccord_app. 2.pdf.

36 The second tranche of CCPL (US$400 million equivalent yen (including US$100 million equivalent yen of Economic and Budget Support Loan)) was provided in December 2009 as the first case of the Hatoyama Initiative, Japan's new policy of providing assistance on climate change to developing countries, announced in September 2009. See MOFA (2009).

37 Unless otherwise indicated, climate data in this section refer to MONRE (2009).

38 See Kitoh *et al.* (2011) for a concise review of climate change projections in Vietnam.

39 Roughly 40 million, or more than half the entire population, is living along the coast and on low-lying islands (World Bank, 2009b).

40 For a detailed assessment of the climate change impact on Vietnam's agriculture sector especially on the crop production sector, see MARD (2010).

41 Vietnam has produced national GHG inventories for 1990, 1993, 1994, 1998, and 2000 under various initiatives (UNDP, 2011). In 2010, JICA began a three-year technical assistance project, "Capacity Building for National Greenhouse Gas Inventory in Viet Nam," which will produce the inventories for 2005 and 2010.

42 The UNFCCC was adopted on May 9, 1992 and entered into force on March 21, 1994. The Kyoto Protocol was adopted on December 11, 1997 and entered into force on February 16, 2005.

43 MONRE Official Document No. 502/BTN-MT-HTQT dated March 24, 2003.
44 MONRE Decision No. 553/QD-BTNMT dated April 29, 2003. The CDM National Executive and Consultative Board was reorganized on 2007 to establish the Steering Committee to implement UNFCCC and KP (MONRE Decision No. 1016/QD-BTNMT dated July 4, 2007).
45 GOV issued the "Guideline for Preparation of Clean Development Mechanism Projects under the Kyoto Protocol" on December 12, 2006 (Circular No. 10/2006/TT-BTNMT) and the "Plan on organization of the implementation of the Kyoto Protocol under the United Nations Framework Convention on Climate Change in the 2007–2010 period" on April 6, 2007 (Decision No. 47/2007/QD-TT). GOV also issued various financial mechanisms to promote CDM investments on August 2, 2007 as Decision 130/2007/QD-TTg.
46 Decision No. 172/2007/QD-TTg dated November 7, 2007.
47 Decision No. 17/2007/QD-TTg dated February 5, 2007.
48 Decision No. 79/2006/QD-TT dated April 14, 2006.
49 Decision No. 177/2007/QD-TTg dated November 20, 2007.
50 Resolution No. 60/2007/NQ-CP dated December 3, 2007.
51 Decision No. 158/2008/QD-TTg of December 2, 2008, approving the National Target Program on Responding to Climate Change. Denmark and UNDP provided technical assistances for its development.
52 Government Office Official Letter No. 282/VPCP-QHQT dated January 13, 2011.
53 Unfortunately, it is not so clear how all these upstream strategies will relate to each other.
54 During the signing ceremony, Ambassador Peter Hansen stressed that one of the main reasons for its decision to support the NTP-RCC was because it showed GOV's commitment to address climate change. See www.ambhanoi.um.dk/en/menu/AboutUs/News/NewsArchives2008/DenmarkFirstInSupportingVietnamsNationalTargetProgramOnClimateChange.htm.
55 The intent was expressed to GOV during JICA's contact mission on February 26–28, 2009 and to the development partners at an unofficial Donor Meeting on Climate Change held on March 10, 2009 and chaired by the UN Resident Coordinator and the Danish Ambassador.
56 The DPL was initially called the Climate Change Program Loan. The name was changed after the Government expressed its concern regarding (1) emphasizing the usage of loan for climate change and (2) limiting the fund resources to loans. This proved to be a valid concern because the Canadian International Development Agency joined SP-RCC with its grant funding.
57 Same as other DPLs, the process is a work in progress. What is described here is the process applied for the first two years, 2009 and 2010.
58 The "perception of priority policy agenda" often comes from the experience of implementing other individual projects. In other words, institutional bottlenecks prohibiting the implementation of projects or formulation of projects are often addressed. This is one of the ways in which DPL synergizes with project-type assistances. Institutional bottlenecks are approached through DPL to enable the effectiveness of other projects.
59 Ideally, this step is replaced by the consolidation of proposals from GOV. It was not done in such way mainly taking into consideration the administrative and coordination work that would have been required from MONRE.
60 JICA provided various technical assistance to assist GOV in realizing some of the policy actions. For example, a study identifying potential mitigation measures in the agriculture sector and institutional bottlenecks of Clean Development Mechanisms were conducted in 2009 and 2010.
61 The only requirement is that it will not be used to procure items in the negative list such as weapons.
62 Government Office, Official Letter No. 8981/VPCP-QHQT.

63 It was agreed in Copenhagen during COP15 that a global fund, the Green Climate Fund, is to be established "to support projects, programme, policies and other activities in developing countries related to mitigation including REDD-plus, adaptation, capacity building, technology development and transfer." It is explicitly stated that this fund is to support actual investments. How this fund is to be financed is, of course, a major issue; however, another critical issue is identifying the eligible portion or the climate change portion of an investment. This is no doubt an area necessary for future global research efforts.

References

ADB. (2005). *Climate Proofing: A Risk-based Approach to Adaptation*, Pacific Studies Series. Manila: ADB.

BAPPENAS (National Development Planning Agency of Republic of Indonesia). (2010a). "Indonesia Climate Change Sectoral Roadmap (ICCSR): Scientific Basis: Analysis and Projection of Temperature and Rainfall," Jakarta, Indonesia.

BAPPENAS (National Development Planning Agency of Republic of Indonesia). (2010b). "Indonesia Climate Change Sectoral Roadmap (ICCSR): Synthesis Report," Jakarta, Indonesia.

Dasgupta, S., B. Laplante, C. Meisner, D. Wheeler, and J. Yan. (2007). "The Impact of Sea Level Rise on Developing Countries: A Comparative Analysis," Policy Research Working Paper 4136. Washington, DC: World Bank.

DNPI (National Climate Change Council). (2010). "Indonesia GHG Abatement Cost Curve," Jakarta, Indonesia.

Dollar, D. and J. Svensson. (2000). "What Explains the Success or Failure of Structural Adjustment Programmes?" *Economic Journal, 110*, 894–917.

IPCC (2007) "Climate Change 2007: Synthesis Report." Contribution of Working Groups I, II and III to the Fourth Assessment Report of the Intergovernmental Panel on Climate Change [Core Writing Team, Pachauri, R.K and Reisinger, A. (eds.)], Geneva, Switzerland: IPCC.

JBIC (Japan Bank for International Cooperation). (2007). "Ex-ante Evaluation of Infrastructure Reform Sector Development Program," Tokyo; JBIC.

JBIC (Japan Bank for International Cooperation). (2008a). "Ex-ante Evaluation of Climate Change Program Loan," Tokyo: JBIC.

JBIC (Japan Bank for International Cooperation). (2008b). "Ex-ante Evaluation of Development Policy Loan IV," Tokyo: JBIC.

JICA (Japan International Cooperation Agency). (2008). "The Study on National Energy Master Plan in Viet Nam: Summary Report," Tokyo: JICA.

JICA (Japan International Cooperation Agency). (2009). "Ex-ante Evaluation of Climate Change Program Loan (II)," Tokyo (www.jica.go.jp/english/operations/evaluation/oda_loan/economic_cooperation/pdf/indo091210_01.pdf).

JICA (Japan International Cooperation Agency). (2010). "Ex-ante Evaluation of Climate Change Program Loan (III) in Indonesia," Tokyo (www.jica.go.jp/english/operations/evaluation/oda_loan/economic_cooperation/pdf/indo100623_01.pdf).

Kitoh, A., S. Kusunoki, Y. Sato, N. Ferdousi, M. Rahman, E.E.S. Makmur, A.L.S. Solis, W. Chaowiwat, and T.D. Trong. (2011). "Climate Change Projection in Some Asian Countries." In Ryo Fujikura and Masato Kawanishi (Eds.), *Climate Change Adaptation and International Development: Making Development Cooperation More Effective*, London and Washington, DC: Earthscan.

MARD (Ministry of Agriculture and Rural Development). (2010). "Adaptation of Crop Production to Climate Change in Vietnam," Hanoi, Vietnam: MARD.

MOE (Ministry of Environment) Republic of Indonesia. (2007). "Climate Variability and Climate Change, and their Implications," Jakarta: MOE.

MOE (Ministry of Environment) Republic of Indonesia. (2010). "Indonesia Second National Communication under the United Nations Framework Convention on Climate Change," Jakarta: MOE.

MOFA (Japanese Ministry of Foreign Affairs). (2007). "Joint Statement by Japan and the Republic of Indonesia on the Enhancement of the Cooperation on Climate Change, Environment and Energy Issues" (www.mofa.go.jp/region/asia-paci/pmv0708/joint.html).

MOFA (Japanese Ministry of Foreign Affairs). (2008). "Climate Change Program Loan (Cool Earth Program Loan)" (www.mofa.go.jp/policy/economy/wef/2008/loan.html).

MOFA (Japanese Ministry of Foreign Affairs). (2009). "Japan's Initiatives to Cope with Global Environmental Problems" (www.mofa.go.jp/policy/environment/pdfs/jp_initiative_pamph.pdf).

MONRE (Ministry of Natural Resources and Environment). (2003). "Viet Nam: Initial National Communication Under the United Nations Framework Convention on Climate Change," Hanoi, Vietnam: MONRE.

MONRE (Ministry of Natural Resources and Environment). (2009). "Climate Change, Sea Level Rise Scenario for Vietnam," Hanoi, Vietnam: MONRE.

MONRE (Ministry of Natural Resources and Environment). (2010). "Viet Nam's Second National Communication to the United Nations Framework Convention on Climate Change,"

Mosley, P., J. Harrigan, and J. Toye. (1991). *Aid and Power: The World Bank and Policy-based Lending Volume I: Analysis and policy proposals*. London: Routledge.

OECF (Overseas Economic Cooperation Fund). (1991). "Issues Related to the World Bank's Approach to Structural Adjustment: Proposal from a Major Partner," *OECF Occasional Paper, 1*, Tokyo, Japan.

Office of Prime Minister. (2007). "Prime Minister's Speech on Global Warming Control," www.kantei.go.jp/foreign/abespeech/2007/05/24speech_e.html.

Parry, M. (2009). "Climate Change is a Development Issue, and Only Sustainable Development can Confront the Challenge" *Climate and Development, 1*, 5–9.

Parry, M., J. Lowe, and C. Hanson. (2009). "Overshoot, Adapt and Recover" *Nature, 458*(7242), 1102–1103.

Shalizi, Z. and F. Lecocq. (2009). "Climate Change and the Economics of Targeted Mitigation in Sectors with Long-Lived Capital," Policy Research Working Paper 5063. Washington, DC: World Bank.

Smith, J.B., S.H. Schneider, M. Oppenheimer, G.W. Yohe, W. Hare, M.D. Mastrandrea, A. Patwardhan, I. Burton, J. Corfee-Morlot, C.H. Magadza, H.M. Füssel, A.B. Pittock, A. Rahman, A. Suarez, and J.P. van Ypersele. (2009). "Assessing Dangerous Climate Change through an Update of the Intergovernmental Panel on Climate Change (IPCC) 'Reasons for Concern'" *Proceedings of the National Academy of Sciences of the United States of America, 106*(11), 4133–4137.

Sudo, Tomonori, Akira Sato, Yuka Murakami, and Mitsunori Motohashi. (2009). "Promotion of Developing Country's Climate Policy Implementation Applying Development Policy Loan," Paper presented at the Asia-Pacific Forum on Low Carbon Economy, Beijing, China, June 17–20, 2009.

Summers, Lawrence and Lant Pritchett. (1993). "The Structural-Adjustment Debate" *American Economic Review, 83*(2), 383–389.

UNDP. (2011). "Climate Change Fact Sheet: Greenhouse gas emissions and options for mitigation in Viet Nam, and the UN's responses," Hanoi, Vietnam: UNDP.

World Bank. (1998). *Assessing Aid: What Works, What Doesn't and Why?* Oxford: Oxford University Press.

World Bank. (2001). *Adjustment Lending Retrospective*, Washington, DC: World Bank.

World Bank. (2006a). *Budget Support as More Effective Aid? Recent Experiences and Emerging Lessons*, Washington, DC: World Bank.

World Bank. (2006b). *Development Policy Lending Retrospective*, Washington DC: World Bank.

World Bank. (2009a). *Investing in a More Sustainable Indonesia: Country Environmental Analysis*, CEA Series, Washington, DC: World Bank.

World Bank. (2009b). *World Development Report 2010: Development and Climate Change*, Washington, DC: World Bank.

World Bank. (2011a). *Climate-Resilient Development in Vietnam: Strategic Directions for the World Bank*, Washington, DC: World Bank.

World Bank. (2011b). *World Development Indicators 2011*, Washington, DC: World Bank.

World Resources Institute. (2011). "Climate Analysis Indicators Tools (CAIT Version 8.0)," http://cait.wri.org.

9

CARBON DIOXIDE REDUCTION EFFECTS OF YEN LOAN COAL-FIRED POWER PLANT PROJECTS

Tomoyo Toyota

Introduction

The Kyoto Protocol requires developed countries to reduce greenhouse gas (GHG) emissions from 2008 to 2012. It has also adopted the Kyoto Mechanisms (Emissions Trading, ET; the Clean Development Mechanism, CDM; and Joint Implementation, JI) as flexible measures that facilitate the achievement of emission goals. Among these mechanisms, only the CDM included developing countries, and it was expected that the mechanism would contribute to effectively reducing GHG emissions. However, the following challenging issues have been identified with regard to projects under the CDM. First, projects require a long time to be authorized. Second, the amount of carbon credits issued will remain unclear until projects are completed. Third, the actual amount of issued credit often turns out to be less than had been assumed. Fourth, the scope of the fields that can be authorized as CDM projects is limited (for example, highly efficient coal-fired power plant projects as well as energy-saving products are not CDM eligible). Fifth, many of the authorized CDM projects actually contribute to the reduction of hydrofluorocarbons (HFCs), but there are a limited number of projects contributing to reduction of carbon dioxide. Last, most authorized projects have been implemented in emerging countries including China and India, while only a few have been implemented in Africa (see Chapter 3).

In order to deal with CDM problems, under Paragraph 11 of the Copenhagen Accord of COP15, the following decision was made:

> In order to enhance action on development and transfer of technology we decide to establish a Technology Mechanism to accelerate technology development and transfer in support of action on adaptation and mitigation that will

be guided by a country-driven approach and be based on national circumstances and priorities.

In response to this, several new initiatives have been launched to push forward the mitigation policy on climate change that includes REDD-plus (Reducing Emissions from Deforestation and Forest Degradation in Developing Countries), a mechanism to prevent deforestation and to facilitate forest conservation (see Chapter 4). Furthermore, as a new initiative, multiple endeavors based on bilateral agreements have been developed, aiming to implement projects for reducing greenhouse gas emissions and measure the results, so that the results from such projects can be certified as actual achievements of greenhouse gas emission reduction. This initiative aims to ensure that projects which had not been authorized as CDM projects can count for credits, including coal-fired power plants or dissemination of energy-saving technology. Under this initiative, projects aimed at infrastructure development as well as private sector driven projects are being considered for inclusion in credit counting.

Japan expresses intensive support for countries and regions that are taking action against climate change. In 2008, then Prime Minister Yasuo Fukuda proposed the "Cool Earth Partnership," which provides further support to developing countries. In 2009, then Prime Minister Yukio Hatoyama proposed the "Hatoyama Initiative" and committed to a target of reducing GHG emissions by 25 percent from 1990 levels by 2020. He also made a commitment to provide financial assistance of 1.75 trillion yen (15 billion US dollars) in combined contributions from the public and private sectors to countries that conclude a Cool Earth Partnership with Japan. Efforts are underway to advance co-benefit type projects, which aim to provide mutual benefits in order to achieve sustainable development as well as address climate change issues in developing countries.

In the area of international environmental cooperation through Japan's Official Development Assistance (ODA) program, projects on promoting energy-saving and diffusing practices for technical efficiency in factories, as well as solar, wind, and geothermal power generation utilizing renewable energy, are receiving recognition as good practices in international cooperation for mitigation. On the other hand, projects related to infrastructure such as road construction or power generators are generally not recognized as methods of international environmental cooperation. Japan has been implementing many ODA projects, a large proportion of which are infrastructure projects: based on the number of loan agreements (L/As), about 25 percent of all projects are in the transportation sector, about 20 percent in the electricity and gas sector, and about 15 percent in the social services sector (such as water and sewage systems, and education). The contribution of infrastructure projects to mitigation has not been evaluated in a quantitative manner, but some projects are making substantial contributions to energy saving and GHG emission reduction. Highly efficient infrastructure facilities established in developing countries through Japan's ODA program will eventually contribute to reducing GHG emissions in these developing countries. In other words, infrastructure projects are of the co-benefit type, which benefits both climate change and development. In this context, this study intends to quantitatively assess the amount of GHG emission reduction accomplished through Japan's ODA program with regard to infrastructure by focusing on coal-fired power projects.

Despite the efforts of the Japanese government, ODA infrastructure projects have never been recognized as supporting the mitigation efforts of developing countries. Japan is among the countries that are keen to be rewarded in carbon credits for projects based on bilateral agreements, and if infrastructure projects are internationally recognized for their mitigation effects, additional support from Japan could be expected. Estimation of the effects of existing projects is thus needed in order to evaluate their CO_2 reduction effects.

This chapter first briefly reviews power generation and associated GHG emissions around the world. It then estimates the effects of GHG emission reduction by means of efficient coal-fired power generation projects funded by yen loans. Coal-fired power plants are among the largest sources of GHG emissions. However, if Japanese ODA can generate electricity with much higher efficiency than the national average in the recipient country, such a project may achieve recognition as a Japanese contribution to GHG emission reduction. Based on this idea, the objective of the present research is to evaluate the GHG emission reduction effects of coal-fired power plants built in ODA projects.

Using several scenarios, this study aims to estimate the GHG emission reduction effects of coal-fired power plants that were developed under an L/A concluded after 1990, which is the year that the Kyoto Protocol sets as a benchmark. Originally, ODA loan infrastructure projects were not regarded as coming under the CDM. Table 9.1 shows the coal-fired power plants that were the subjects of this study. Many yen loan funded power plants that have been in service for a relatively long time are operating in China and India. Up to 2010, such coal-fired power plants have been generating power with a total scale of approximately 7,200 MW in four countries: China, India, Malaysia, and Vietnam. This study will estimate the cumulative volume of the GHG emission reduction effects of ten power plants in these four countries, as shown in Table 9.1.

Electricity, energy structure, and GHG emissions

Electricity and GHG emissions around the world

Global CO_2 emissions were about 30 billion tons of CO_2 per year in 2007, which was double the 1970 level. Annual CO_2 emissions of non-OECD countries exceeded Organization for Economic Cooperation and Development (OECD) countries in 2003, and this difference has been increasing (Figure 9.1, left). Figure 9.1 indicates that total GHG emissions have been rising due to the increase in the power generation sector. "Main Activity Producer Electricity and Heat" accounts for the greatest proportion, with "Transportation" and "Manufacturing Industries and Construction" in second place.

TABLE 9.1 Yen loan projects examined in this study

Country	*Project name*	*Number of L/As*	*Project cost total billion yen: nominal*	*Within an ODA loan billion yen: nominal*	*Project period*	*Generation capacity MW*
China	Hubei Ezhou thermal power plants projects	3	61.9	31.7	1992–2001	600
	Jiangxi Jiujiang thermal power plant project	2	48.9	25.7	1995–2003	700
	Sanhe thermal power plant project	2	54.5	23.0	1995–2002	700
	Shanxi Hejin thermal power plant project	2	44.3	22.8	1995–2002	700
India	Anpara B thermal power station construction project	5	140.2	118.4	1984–2001	1,000
	Bakreswar thermal power station project/unit 3 extension construction project	4	93.7	77.3	1994–2004	640
	Kothagudem'A' thermal power station rehabilitation project	1	6.2	5.1	1995–2000	240
	Simhadri thermal power station project	4	91.3		1997–2004	1,000
Malaysia	Port Klang power station project	2	84.0	54.8	1992–2001	1,000
Vietnam	Pha Lai thermal power plant project	4	70.6	65.1	1995–2003	600

Source: JICA.

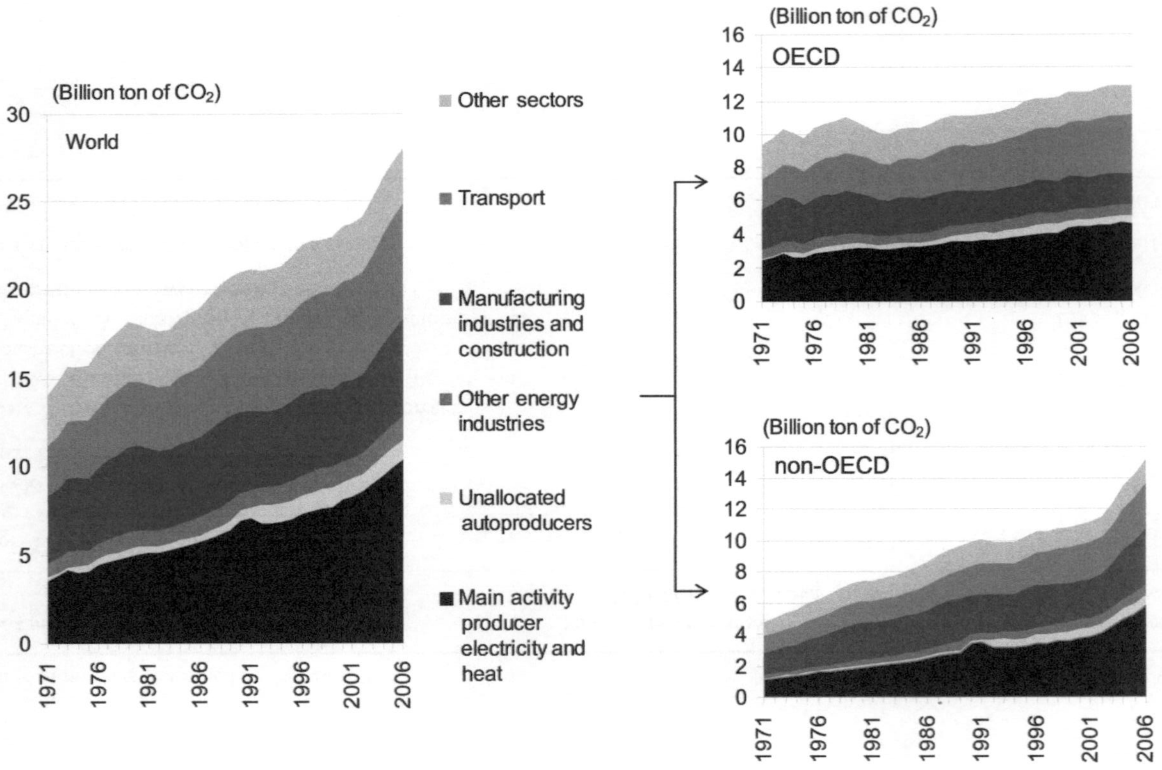

FIGURE 9.1 GHG emissions by sector in OECD and non-OECD countries (source: IEA, 2008a).

Table 9.2 shows basic economic, environmental, and energy information among OECD and non-OECD countries. CO_2 emissions per capita in non-OECD countries remain about one-fifth those of OECD countries. Further increases in CO_2 emissions are anticipated as industrial activities expand with continuing economic growth. In 2007, average CO_2 emissions per capita in non-OECD countries were 2.4 tons per year. When this reaches 11.1 tons per year, the present level among OECD countries, global CO_2 emissions will reach 42.8 billion tons per year, equivalent to 1.4 times the present amount of emissions.

The major contributing factor in the increase in CO_2 emissions in recent years has been the increase in emissions from the electric power sector in non-OECD countries. From 2000 to 2007, the amount of power generation per capita in non-OECD countries increased 1.45 times, while that in OECD countries increased 1.06 times. Although electric power generation per capita in non-OECD countries still remains less than one-fifth that of OECD countries, the quantity of GHG emissions from the electric power sector in non-OECD countries will further increase in the future.

As shown in Table 9.2, energy intensity in OECD countries improved from 0.23 in 1990 to 0.17 in 2007, and that of non-OECD countries also improved from 0.36 to 0.27 during the same period. The carbon intensity of an energy supply is the CO_2 emission per unit of energy used, which directly links changes in CO_2 emissions to changes in energy usage, and is dependent on the energy source involved. Coal is the most carbon-intensive fuel, followed by oil and natural gas. Over time, declining carbon intensity can offset increasing energy consumption to some extent. Carbon intensity has been tending to improve in both OECD and non-OECD countries.

The global electrification ratio was 78.2 percent in 2008 (93.4 percent in urban areas and 63.2 percent in rural areas), and the population without electricity was 1,456 million (Table 9.3). Most non-electrified areas are located in rural areas of developing countries; the electrification rates in rural areas of South Asia and Sub-Saharan Africa are especially low, at 51.2 percent and 14.3 percent, respectively. Further increases in GHG emissions from the electric power generation sector can be expected due to the increased demand for electric power and the expansion of the electric power supply in non-electrified areas in developing countries.

Power generation, energy efficiency, and generation resources around the world

Figure 9.2 shows power generation by source from 1990 to 2006 among OECD and non-OECD countries. Coal is the most widely used fuel for power generation, and accounts for more than 40 percent of the world's power generation. Although the share of coal used in the power generation sector is decreasing in the OECD countries, and despite the fact that the share of natural gas is tending to increase, coal still supplies the biggest single generation method, at about 40 percent. The share of nuclear power plants is the next highest after coal, and accounts for more

TABLE 9.2 Basic data on energy, economy, and emissions by region, 1970–2007

	1970	*1980*	*1990*	*2000*	*2007*		*1970*	*1980*	*1990*	*2000*	*2007*
Population (million people)						*Electricity production per capita (MWh/year)*					
OECD	785	887	983	1,063	1,150	OECD	4.11	5.78	7.17	8.44	8.92
Non-OECD	2,247	2,800	3,462	4,216	4,935	Non-OECD	n.a.	0.73	1.00	1.16	1.67
East Asia and Pacific	1,049	1,291	1,557	1,817	2,038	East Asia and Pacific	n.a.	0.75	1.13	1.74	2.80
South Asia	564	714	906	1,129	1,362	South Asia	n.a.	0.15	0.30	0.48	0.62
Europe and Central Asia	668	739	796	844	864	Europe and Central Asia	n.a.	4.50	5.40	5.39	5.98
Middle East and North Africa	105	138	186	255	317	Middle East and North Africa	n.a.	0.73	1.28	1.93	2.61
Sub-Saharan Africa	229	294	387	515	672	Sub-Saharan Africa	n.a.	0.42	0.51	0.49	0.54
North America	199	226	252	277	313	North America	8.10	11.12	13.28	14.80	14.83
Latin America and Caribbean	218	284	361	441	519	Latin America and Caribbean	n.a.	1.03	1.42	1.93	2.28
Income level (GDP, PPP) (constant 2005 int. 1,000 $) per capita						*Energy intensity (TOE of energy use per 1,000 2005 int. $ of GDP, PPP)*					
OECD	n.a.	n.a.	18.0	22.6	27.5	OECD	n.a.	n.a.	0.23	0.19	0.17
Non-OECD	n.a.	n.a.	2.5	2.8	3.3	Non-OECD	n.a.	n.a.	0.36	0.34	0.27
East Asia and Pacific	n.a.	n.a.	2.5	3.8	5.4	East Asia and Pacific	n.a.	n.a.	0.33	0.27	0.22
South Asia	n.a.	n.a.	0.9	1.2	1.7	South Asia	n.a.	n.a.	0.31	0.28	0.24
Europe and Central Asia	n.a.	n.a.	13.3	15.8	17.3	Europe and Central Asia	n.a.	n.a.	n.a.	0.24	0.18
Middle East and North Africa	n.a.	n.a.	7.0	6.2	7.3	Middle East and North Africa	n.a.	n.a.	0.13	0.19	0.21
Sub-Saharan Africa	n.a.	n.a.	1.8	1.6	1.5	Sub-Saharan Africa	n.a.	n.a.	0.34	0.38	0.38
North America	n.a.	n.a.	25.3	31.4	38.9	North America	n.a.	n.a.	0.31	0.24	0.21
Latin America and Caribbean	n.a.	n.a.	7.6	7.1	8.3	Latin America and Caribbean	n.a.	n.a.	0.14	0.15	0.14

CO_2 emission per capita (ton/year)						*Carbon intensity (ton of CO_2 per TOE of energy use)*					
OECD	7.4	10.6	11.2	10.8	11.1	OECD	3.00	2.86	2.71	2.55	2.41
Non-OECD	1.6	2.0	2.4	2.6	2.4	Non-OECD	n.a.	n.a.	2.78	2.73	2.69
East Asia and Pacific	1.2	1.6	2.1	2.8	3.2	East Asia and Pacific	n.a.	n.a.	2.60	2.73	2.69
South Asia	0.2	0.3	0.4	0.7	1.0	South Asia	n.a.	n.a.	1.56	2.01	2.38
Europe and Central Asia	5.6	8.2	10.2	9.7	7.4	Europe and Central Asia	n.a.	n.a.	n.a.	2.61	2.33
Middle East and North Africa	0.9	2.3	3.5	3.8	4.8	Middle East and North Africa	n.a.	n.a.	3.81	3.20	3.11
Sub-Saharan Africa	0.5	0.8	1.0	0.9	0.8	Sub-Saharan Africa	n.a.	n.a.	1.61	1.50	1.43
North America	15.7	21.0	20.4	19.1	20.0	North America	2.84	2.81	2.58	2.50	2.48
Latin America and Caribbean	1.4	1.8	2.5	2.4	2.6	Latin America and Caribbean	n.a.	n.a.	2.31	2.26	2.22

Source: World Bank (2010), IEA (2008a, b).

TABLE 9.3 Electrification rates and population without electricity in 2008

	Electrification rate		*Population without electricity*	
	Overall, %	*Urban, %*	*Rural, %*	*Millions*
World	78.9	93.6	65.1	1,441
Developing countries	73.0	90.7	60.2	1,438
Developing Asia	78.1	93.9	68.8	799
China and East Asia	90.8	96.4	86.5	186
South Asia	62.2	89.1	51.2	612
Africa	41.9	68.9	25.0	587
North Africa	99.0	99.6	98.4	2
Sub-Sahara Africa	30.5	59.9	14.3	585

Source: IEA (2010).

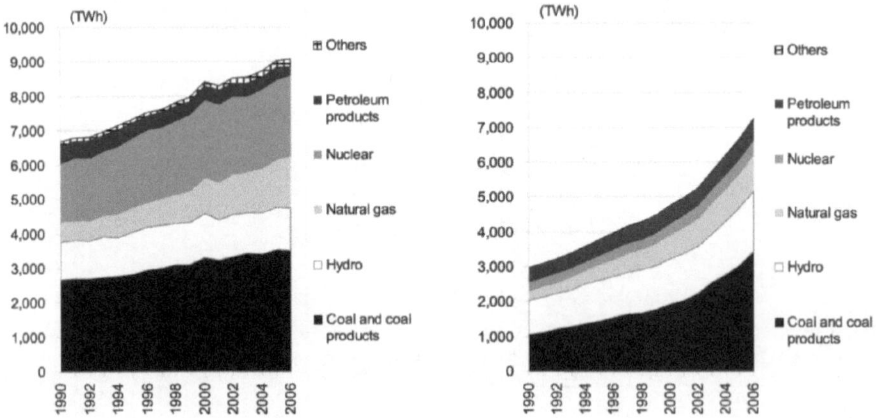

FIGURE 9.2 Power generation structure change (left: OECD countries; right: non-OECD countries) (source: IEA, 2008a).

than one-quarter of the power produced by OECD countries. The share of hydroelectric power decreased from 16.3 percent in 1990 to 13.5 percent in 2007, and natural gas expanded from 8.7 to 16.7 percent.

Coal is the also largest source for power generation in non-OECD countries, and about 50 percent of their electricity was generated by coal-fired power plants in 2007. The amount of power generated by coal-fired power plants is increasing, and the amount of electricity produced has increased about 3.2 times from 1990 to 2006. The share of hydroelectric power decreased from 32.5 percent in 1990 to 23.9 percent in 2007, and that of natural gas became four times greater, from 9 percent to 14.5 percent. Non-OECD countries are heavily dependent on thermal power plants for power generation, and especially coal fueled power generation.

Coal has been the world's most widely used source of fuel for power generation because of its cheap price. Figure 9.3 shows the worldwide emission factors (kg CO_2/kWh) by fuel category in the world's thermal power plants. According to

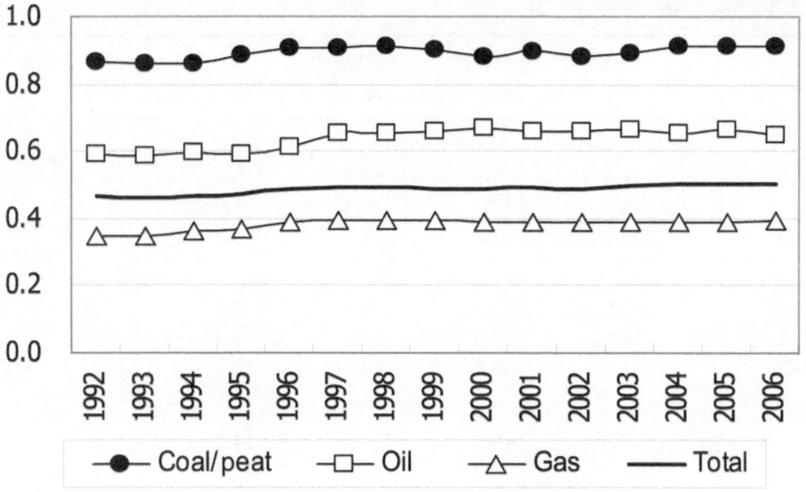

FIGURE 9.3 CO_2 emissions per KWh of electricity and heat output (kg CO_2 per kWh) (source: IEA, 2008a, b).

International Energy Outlook 2010 (EIA 2010), if the price system does not fluctuate, coal will continue to account for the greatest proportion of electric power production, and net world coal-fired generation is expected to nearly double from 2007 to 2035. It is thus important to adopt appropriate mitigation policies for the power generation sector.

The amount of CO_2 emission per kilowatt hour of power generation (called the emission factor) differs significantly depending on the fuel category, the electricity generation method, the technology to be used, and so on. The emission factor is determined by the ratio of the thermal energy (heat) of the input fossil fuel to the amount of power generation (thermal efficiency). The IEA uses two kinds of efficiency data for electricity: thermal efficiency to amount of power generation (net generating efficiency), and thermal efficiency to amount of power transmission (net generating efficiency), which takes account of power transmission loss. In this study, the former kind of thermal efficiency will be used, due to the available data. Table 9.4 shows the thermal efficiency according to fuel type. From 1990 to 2006, both OECD and non-OECD countries improved their thermal efficiency. However, for the world as a whole, the thermal efficiency of coal fuel slightly decreased, due to the increased share in this period of non-OECD countries with lower thermal efficiency.

Power generation and energy structure in China, India, Malaysia, and Vietnam

Energy supply structure in the power generation sector

Figure 9.4 shows changes in the energy supply structure of the power sector in China, India, Malaysia, and Vietnam; in all four countries, the total power generation

TABLE 9.4 Thermal efficiency of fuels

	1990 %	*2000* %	*2006* %	*Change rate (1990–2006)* %
Coal				
World	34.6	34.9	34.4	–0.4
OECD	36.6	37.1	37.3	1.9
Non-OECD	30.3	31.8	31.9	5.3
Petroleum products				
World	37.0	39.0	39.0	5.4
OECD	38.9	39.9	37.6	–3.3
Non-OECD	34.9	38.3	39.6	13.7
Natural gas				
World	37.3	40.2	42.5	14.2
OECD	39.2	43.2	47.3	20.6
Non-OECD	33.7	36.3	37.2	10.4

Source: IEA (2008a, b).

has been increasing rapidly since the mid-1980s. China mainly uses coal, which accounts for 80 percent of its power generation, and its share is increasing. It is estimated that China has more than 10 percent of the world's coal resources, and it uses coal as its primary resource in supplying power. India also mainly uses coal as fuel, which accounts for 70 percent of its power generation. Although its coal ratio was higher in the past, Malaysia now uses natural gas for more than half of its electrical power. In Vietnam, the proportion of coal used is the lowest in these four countries: hydroelectric power generation accounts for almost half of its power, and the share of natural gas has also been increasing recently (IEA 2008a).

In 1970, India accounted for only 1 percent of the world's power generation, but its share has been expanding since the 1990s and reached 4 percent in 2006. China's share has rapidly increased since the middle of the 1980s, and expanded from 3 percent of the world's power generation in the early 1970s to 10 percent in 2000. In addition, the power demand in China has increased by an average of 10 percent a year since 2000, and its overall share of global production expanded to almost 18 percent of the world supply (in 2006). Malaysia and Vietnam together account for less than 1 percent of global electricity production: 0.5 percent and 0.3 percent, respectively, in 2006 (IEA 2008a).

Energy efficiency

Figure 9.6 shows the average thermal efficiency of power plants from 2000 to 2006; data for Japan and the world average are included for reference. The thermal efficiency of coal is relatively low compared to oil and natural gas, and the quantity of GHG emissions per unit of power generation is large. Japan is one of the world's most efficient countries in terms of its electricity generation efficiency. Energy

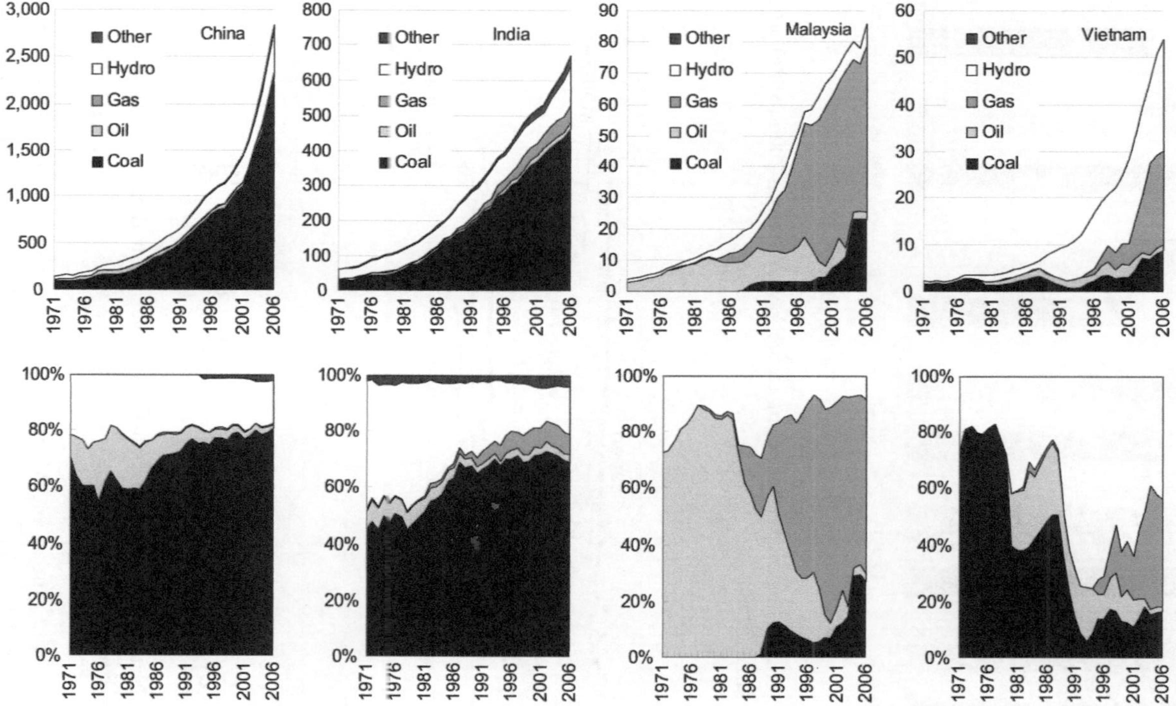

FIGURE 9.4 Power generation change according to fuel (top) and share of power generation structure (bottom) (source: IEA, 2008a, b).

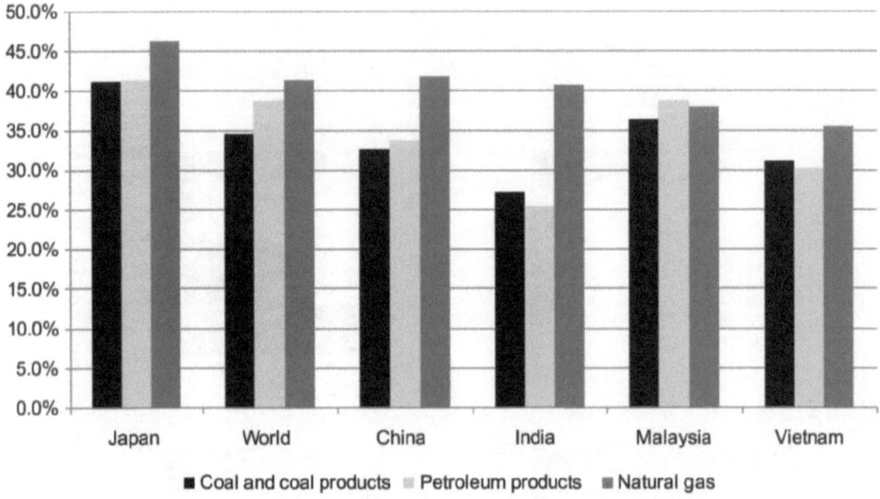

FIGURE 9.5 Thermal efficiency according to fuel (average from 2000 to 2007) (source: IEA, 2008a).

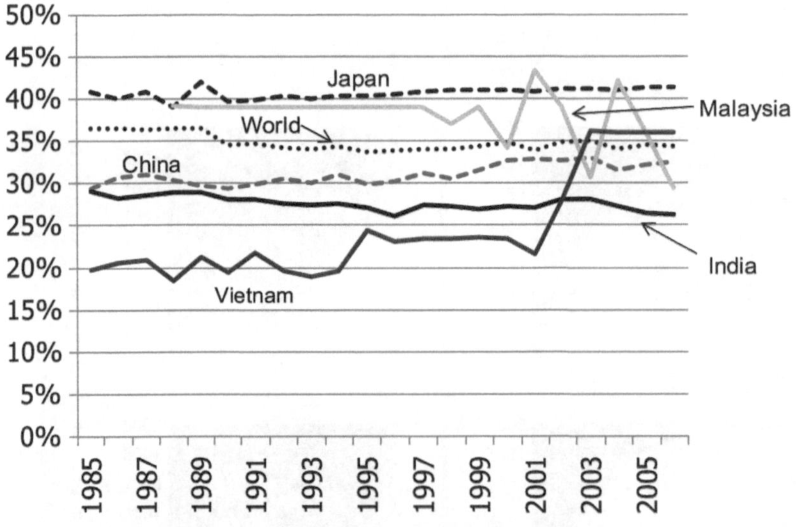

FIGURE 9.6 National averages of coal-fired power plant thermal efficiencies (source: IEA, 2008a).

efficiency in China, India, and Vietnam are lower than the global average, which implies there is plenty room for improvement. Figure 9.6 shows changes in the thermal efficiency of coal-fired power plants by country. Japan's efficiency has been more than 40 percent since 1985; the world average was around 35 percent over the same period.

Methodology

The GHG emission reduction effect is calculated as the emission gap between the baseline and coal-fired power plants built in ODA projects. Referring to methods used in the CDM, this study prepared four baseline scenarios, namely reference scenarios 1 to 4 (RF1 to RF4).

Cases with reference to CDM methodologies

Reference scenario RF1 was prepared according to the CDM methodology ACM0013: "Consolidated baseline and monitoring methodology for new grid connected fossil fuel-fired power plants using a less GHG intensive technology." In order to be applicable, the methodology states that the following conditions are required (source: CDM methodology ACM0013):

- the project activity is the construction and operation of a new fossil fuel-fired grid-connected electricity generation plant that uses a more efficient power generation technology than what would otherwise be used with the given fossil fuel;
- the project activity is not a co-generation power plant;
- data on fuel consumption and electricity generation of recently constructed power plants are available;
- the identified baseline fuel is used in more than 50 percent of total generation by utilities in the geographical area within the host country, as defined later in the methodology, or in the entire host country. To demonstrate this applicability condition data for the latest three years shall be used. Maximum value of same fossil fuel generation estimated for three years should be greater than 50 percent.

For the above reasons, ACM0013 is only applicable with newly constructed power plants: the rehabilitation of existing facilities is not included, even though highly effective technology may be adopted. In addition, this methodology can only be used in a region where more than 50 percent of total electricity is generated using the subject fossil fuel. Under these constraints, some of the conditions do not apply in the case of the subject projects, but it is worth using this CDM as a reference in estimating the amount of GHG emission reduction.

The GHG emission reduction amount is calculated as the difference between the project emission (PE_y) and baseline emission (BE_y) amounts:

$$ER_y = BE_y - PE_y, \tag{1}$$

where ER_y is emission reduction in year y (ton-CO_2), BE_y is baseline emission in year y (ton-CO_2), and PE_y is project emission in year y (ton-CO_2).

PE_y is CO_2 emission from a coal-fired power plant built with yen loan funding. BE_y is calculated by multiplying the electricity generated in the project plant by a baseline CO_2 emission factor as follows:

$$BE_y = EG_{PJ,y} \times EF_{BL,CO2,y}, \tag{2}$$

where BE_y is baseline emission in year y (ton-CO_2), $EG_{PJ,y}$ is the net quantity of electricity generated by the project plant in year y (MWh), and $EF_{BL,CO2,y}$ is the baseline emission factor in year y (ton-CO_2/MWh).

In RF1, $EF_{BL,CO2,y}$ is determined according to the build margin (BM) in the CDM methodology, which assumes that another conventional coal-fired power plant would be built instead of the implemented project to generate the same amount of power. The emission factor used in the BM is assumed to be equal to the national average of coal-fired power plants in the respective country. The amount of CO_2 reduction with the BM ($ER_{BM,y}$) can be calculated as follows:

$$ER_{BM,y} = EG_{PJ,y} \times (EF_{BL\_coal,CO2,y} - EF_{PE\_coal,CO2,y}), \tag{3}$$

where $ER_{BM,y}$ is the BM emission reduction effect (ton-CO_2), $EG_{PJ,y}$ is the net quantity of electricity generated by the project plant in year y (MWh), $EF_{BL\_coal,CO2,y}$ is the average emission factor in the applicable country (ton-CO_2/MWh), and $EF_{PE\_coal,CO2,y}$ is the emission factor of a coal-fired power plant funded by a yen loan (ton-CO_2/MWh).

In RF2, $EF_{BL,CO2,y}$ is determined according to the operating margin (OM) in the CDM methodology, which assumes that the operation rate of existing power plants would increase in the absence of the implemented project. Because it is unrealistic to assume that the operation rate of renewable energy sources could increase, the emission factor is determined according to the average of all thermal power plants excluding renewable energy sources. The amount of CO_2 reduction with the OM ($ER_{OM,y}$) can be calculated as follows:

$$ER_{OM,y} = EG_{PJ,y} \times (EF_{OL\_thermal,CO2,y} - EF_{PE\_coal,CO2,y}), \tag{4}$$

where $ER_{OM,y}$ is the CO_2 emission reduction effect of the OM (ton-CO_2), $EG_{PJ,y}$ is the net quantity of electricity generated in the project plant in year y (MWh), and $EF_{OL\_thermal,CO2,y}$ is the average emission factor of thermal power plants (ton-CO_2/MWh).

$EF_{BL\_thermal,CO2,y}$ is calculated assuming that the operation rates of existing thermal power plants (coal, gas, and oil) rise equally. The average emission factor for thermal power plants include gas- and oil-fired plants, which have higher thermal efficiency than coal. The national average thermal efficiency of thermal power plants can be higher than that of coal-fired power plants built in a yen loan project if the country depends on gas and/or oil for power generation. In that case, the effect of the project becomes negative: the emission amount of the project increases above the baseline.

Taking new power plant construction into consideration

Under the RF1 and RF2 scenarios, the impact of new projects is likely to be overestimated because the projects are compared with a national average, which includes old and low-efficiency power plants. Reference scenario 3 (RF3) and reference scenario 4 (RF4) are prepared to avoid such overestimation.

RF3 is based on RF1. Instead of the national average thermal efficiency, that of newly built coal-fired power plants in the country during a range of years from y to $y-1$ is adopted as a reference. RF3 is determined as follows:

$$ER_{BMnew,\,y} = EG_{PJ,y} \times (EF_{BL\_coalnew,CO2,y} - EF_{PE\_coal,CO2,y}), \tag{5}$$

where $EF_{Bmnew,y}$ is the reduction amount in RF3 (ton-CO_2), and $EF_{BL\_coalnew,CO_2,y}$ is the emission factor of new coal-fired plants that started operations in y, calculated as

$$EF_{BL\_coalnew,CO_2,y} = EF_{BL\_coalnew,CO_2,y} - EF_{BL\_coalnew,CO_2,y\text{-}1}.$$

RF4 is based on RF2. It further assumes that the thermal efficiency of the newly constructed thermal power plants in y can be calculated from the difference between power generation by thermal power plants and the fuel input from $y-1$ to y. RF4 is determined as follows:

$$ER_{OMnew,\,y} = EG_{PJ,y} \times (EF_{BL\_thermal,CO2,y} - EF_{PE\_coal,CO2,y}) \tag{6}$$

where $EF_{OMnew,y}$ is the CO_2 reduction amount in RE4 (ton-CO_2), with

$$EF_{OM\_thermalnew,CO_2,y} = EF_{OM\_thermalnew,CO_2,y} - EF_{OM\_thermalnew,CO_2,y\text{-}1}. \tag{7}$$

Difference in energy efficiency between baseline and project

The emission factor of a coal-fired power plant built in an ODA project is usually lower than the national average for coal-fired power plants in each country. Table 9.5 summarizes the average thermal efficiency of ODA project plants and RF1 to RF4 by country. The thermal efficiency of the project in China is comparable to that of power plants built in Japan and is the highest among the four countries, followed by Vietnam, Malaysia, and India. In terms of RF1, the energy-saving effects in China and India are high because the efficiencies of the projects are more than 30 percent higher than the baseline. The thermal efficiencies of the RF2 baseline are better than RF1 because RF2 includes gas and oil, which have high efficiency. The difference is not significant in China and India because of the low share of gas and oil in power generation in those countries. The thermal efficiency for Malaysia is lower for RF2 than for RF1 because of the high share of oil in Malaysia's power generation. The thermal efficiencies in RF3 and RF4 are slightly higher than RF1

TABLE 9.5 Thermal efficiency in reference scenarios

	Project average %	*National average (1990–2006)*			
		RF1 (existing coal power plant) %	*RF2 (existing fossil fuel power plant) %*	*RF3 (new coal power plant) %*	*RF4 (new fossil fuel power plant) %*
China	42.6	32.4	32.5	33.6	38.0
India	35.0	27.0	27.7	22.9	27.0
Malaysia	36.5	32.8	36.1	35.1	34.4
Vietnam	38.3	36.1	36.4	36.7	36.4
Japan (reference)	n.a.	40.7	42.8	39.9	36.0

Source: author.

and RF2 because newly constructed power plants are included in the RF3 and RF4 calculations. The GHG emission reduction effect is calculated as follows:

$$ER_{i,y} = EF_{i,y} \times GAP_{i,y} \times EG_{i,y} \quad (8)$$

where $ER_{i,y}$ is the CO_2 reduction amount in i project plant in year y (ton-CO_2/year), $GAP_{i,y}$ is the difference between the emission factors of the project and the baseline, and $EG_{i,y}$ is the net quantity of electricity generated in the i project plant in year y (MWh/year).

Results

Table 9.6 shows the results of the GHG emission reduction effect estimation under the four scenarios. Where more than one power plant is involved in the study (i.e., four in China, and five in India) the effects are the sum of the effects of the plants. Negative figures indicate when the thermal efficiency of a project is lower than the baseline, which means the GHG emissions from the project are larger than the baseline. Figures in parentheses show ratios of GHG emission reduction to the total emissions by the electric power sectors of the respective countries.

Under RF1, most reduction amounts are positive, suggesting that the projects contributed to GHG emission reduction. These reductions are significant in India and China, where the energy efficiencies of the projects are high and the amount of power generation is large. Cumulative GHG emission reduction until 2007 is positive in every case in RF1; however, under RF2, the reduction effect is positive in China and India, but negative in Malaysia and Vietnam. In the latter countries, the share of natural gas used in power generation is large, and the average efficiency of thermal power plants is higher than that of the projects, which implies that GHG emissions increased above the baseline values.

TABLE 9.6 Estimated GHG emission reduction effects

	China	India	Malaysia	Vietnam
RF1				
1994	n.a.	1,132 (0.4%)	n.a	n.a.
1995	n.a.	1,294 (0.4%)	n.a.	n.a.
1996	n.a.	1,582 (0.4%)	n.a.	n.a.
1997	n.a.	1,227 (0.3%)	n.a.	n.a.
1998	n.a.	1,333 (0.3%)	n.a.	n.a.
1999	289 (0.0%)	1,414 (0.3%)	n.a.	n.a.
2000	883 (0.1%)	1,627 (0.3%)	n.a.	n.a.
2001	1,733 (0.1%)	2,075 (0.4%)	n.a.	n.a.
2002	1,677 (0.1%)	1,806 (0.4%)	–217 (–0.5%)	n.a.
2003	2,269 (0.1%)	1,931 (0.4%)	716 (1.8%)	41 (0.3%)
2004	3,533 (0.2%)	3,253 (0.6%)	–673 (–1.6%)	–2 (–0.0%)
2005	3,054 (0.1%)	4,603 (0.8%)	75 (0.1%)	25 (0.1%)
2006	3,051 (0.1%)	4,915 (0.8%)	1,115 (2.0%)	10 (0.1%)
2007	3,159 (0.1%)	4,915 (0.8%)	973 (1.6%)	–55 (–0.2%)
Total	19,649	33,108	1,990	19
RF2				
1994	n.a.	1,010 (0.3%)	n.a.	n.a.
1995	n.a.	1,110 (0.3%)	n.a.	n.a.
1996	n.a.	1,336 (0.3%)	n.a.	n.a.
1997	n.a.	1,183 (0.3%)	n.a.	n.a.
1998	n.a.	1,143 (0.3%)	n.a.	n.a.
1999	271 (0.0%)	1,051 (0.2%)	n.a.	n.a.
2000	851 (0.1%)	1,357 (0.3%)	n.a.	n.a.
2001	1,653 (0.1%)	1,709 (0.3%)	n.a.	n.a.
2002	1,610 (0.1%)	1,464 (0.3%)	57 (0.1%)	n.a.
2003	2,184 (0.1%)	1,495 (0.3%)	–312 (–0.8%)	–25 (–0.2%)
2004	3,366 (0.2%)	2,556 (0.5%)	–535 (–1.3%)	–18 (–0.1%)
2005	2,933 (0.1%)	3,639 (0.6%)	215 (0.4%)	–1 (–0.0%)
2006	2,942 (0.1%)	3,857 (0.6%)	224 (0.4%)	–11 (–0.1%)
2007	3,047 (0.1%)	3,857 (0.6%)	208 (0.3%)	–58 (–0.3%)
Total	18,856	26,769	–142	–114
RF3				
1994	n.a.	771 (0.3%)	n.a.	n.a.
1995	n.a.	2,029 (0.6%)	n.a.	n.a.
1996	n.a.	4,358 (1.1%)	n.a.	n.a.
1997	n.a.	–9,577 (–2.4%)	n.a.	n.a.
1998	n.a.	2,904 (0.7%)	n.a.	n.a.
1999	–387 (–0.0%)	1,983 (0.4%)	n.a.	n.a.
2000	–1,144 (–0.1%)	833 (0.2%)	n.a.	n.a.
2001	1,132 (0.1%)	3,169 (0.6%)	n.a.	n.a.
2002	1,884 (0.1%)	–35,081 (–7.0%)	1,115 (2.6%)	n.a.

continued

TABLE 9.6 Continued

	China	*India*	*Malaysia*	*Vietnam*
2003	1,828 (0.1%)	2,531 (0.5%)	2,416 (6.1%)	41 (0.3%)
2004	6,033 (0.3%)	8,782 (1.6%)	–673 (–1.6%)	–407 (–2.3%)
2005	1,368 (0.1%)	12,731 (2.2%)	1,104 (2.1%)	33 (0.2%)
2006	2,219 (0.1%)	6,621 (1.1%)	1,389 (2.4%)	–6 (–0.0%)
2007	2,304 (0.1%)	6,621 (1.0%)	1,213 (2.0%)	–71 (–0.3%)
Total	15,236	8,675	6,565	–410
RF4				
1994	n.a.	–465 (–0.2%)	n.a.	n.a.
1995	n.a.	1,523 (0.4%)	n.a.	n.a.
1996	n.a.	3,200 (0.8%)	n.a.	n.a.
1997	n.a.	–2,959 (–0.7%)	n.a.	n.a.
1998	n.a.	–34 (–0.0%)	n.a.	n.a.
1999	–421 (–0.0%)	–170 (–0.0%)	n.a.	n.a.
2000	–1,177 (–0.1%)	2,502 (0.5%)	n.a.	n.a.
2001	1,088 (0.1%)	2,264 (0.5%)	n.a.	n.a.
2002	1,795 (0.1%)	–15,320 (–3.1%)	919 (2.2%)	n.a.
2003	1,755 (0.1%)	1,067 (0.2%)	–312 (–0.8%)	–25 (–0.2%)
2004	5,572 (0.3%)	6,973 (1.2%)	–535 (–1.3%)	109 (0.6%)
2005	1,236 (0.1%)	11,643 (2.0%)	215 (0.4%)	44 (0.2%)
2006	2,117 (0.1%)	4,829 (0.8%)	140 (0.2%)	–51 (–0.3%)
2007	2,198 (0.1%)	4,829 (0.7%)	130 (0.2%)	–97 (–0.4%)
Total	14,162	19,882	558	–21

Source: author.

RF3 and RF4 were prepared in order to avoid overestimation of the effect of the projects. As expected, the estimated reduction effects in China and Vietnam are smaller than those calculated under RF1 and RF2. However, they are larger than the RF1 and RF2 calculations in India and Malaysia, because the energy efficiency of their newly constructed power plants is relatively low.

Table 9.7 shows the cumulative effect on CO_2 emissions of all of the coal-fired power plants from 2000, when five of them actually started operations, to 2007, and also includes cumulative CO_2 emissions in Japan. The cumulative effect of emission reductions through the projects is equivalent to about 0.8 percent.

Conclusion

This study demonstrated that highly efficient coal-fired power plant projects funded by yen loans substantially contributed to CO_2 reduction in Asian countries. Four scenarios were prepared based on a CDM methodology in order to set emission baselines. The cumulative amount of GHG emission reduction in ten Japanese

TABLE 9.7 Cumulative CO_2 emissions in Japan and cumulative CO_2 reduction effects from 2000 to 2007 (unit: kt-CO_2)

	Japan★ kt-CO_2	*Cumulative effect of ODA on GHG reduction*							
		RF1		*RF2*		*RF3*		*RF4*	
		kt-CO_2	*%*	*kt-CO_2*	*%*	*kt-CO_2*	*%*	*kt-CO_2*	*%*
2000	1,228,788	2,510	(0.2%)	2,208	(0.2%)	–311	(0.%)	1,325	(0.1%)
2001	2,450,227	6,318	(0.3%)	5,570	(0.2%)	3,990	(0.2%)	4,678	(0.2%)
2002	3,686,922	9,584	(0.3%)	8,700	(0.2%)	–28,092	–(0.8%)	–7,929	–(0.2%)
2003	4,945,158	14,541	(0.3%)	12,042	(0.2%)	–25,828	–(0.5%)	–8,769	–(0.2%)
2004	6,208,655	20,653	(0.3%)	17,411	(0.3%)	–14,789	–(0.2%)	–4,540	–(0.1%)
2005	7,450,066	28,410	(0.4%)	24,198	(0.3%)	3,911	(0.1%)	11,864	(0.2%)
2006	8,685,032	37,502	(0.4%)	31,210	(0.4%)	18,496	(0.2%)	18,898	(0.2%)
2007	9,938,549	46,494	(0.5%)	38,264	(0.4%)	32,372	(0.3%)	25,959	(0.3%)

Source: author.

Notes

★ WDI (2010). Data in parentheses: means ratio of cumulative GHG emission in Japan.

ODA projects implemented in China, India, and Malaysia from 2000 to 2007 was in a range of 41–75 million ton-CO_2, which is equivalent to 0.4–0.8 percent of Japan's total emissions during the same period.

Coal is the most "dirty" fuel in terms of CO_2 emissions. However, carbon dioxide capture and storage (CCS) technology is not yet available for practical use, either technologically or economically. Highly efficient coal-fired power generation will remain a viable option in developing countries as long as coal is a cheap and easily available fuel. With the exception of Vietnam, where hydro provides a substantial share of the nation's power generation, establishing highly efficient coal-fired power plants will substantially reduce carbon intensities. If China, India, and Malaysia could have replaced their power plants with efficient plants, the GHG emission reduction effect in 2007 would have been 557.5, 113.8, and 0.06 million tons, respectively. The total amount of this potential reduction, 671.4 million tons, is equivalent to 53.6 percent of the total GHG emissions in Japan.

Table 9.8 shows that 62 yen loan projects, the L/As of which were concluded after 1990, have been implemented in the electric power generation sectors of 17 countries. Considering the emission reduction effect of only ten of these projects, the total effect of yen loan projects can be expected to be significant compared with Japan's own emissions and its reduction obligations under the Kyoto Protocol. This contribution did not result from authorized CDM projects, but can be recognized as a case of substantial action for mitigation efforts using yen loans.

TABLE 9.8 Yen loan projects in the power generation sector

Power-generating capacity (MW)	*Thermal power plants*						*Renewable power plants*			
	Coal	*Oil*	*Gas*	*Both natural gas and crude oil*	*Combined cycle (oil)*	*Combined cycle (natural gas)*	*Hydroelectric*	*Pumped-up*	*Geothermal*	*Wind*
Azerbaijan	–	–	–	–	–	400	–	–	–	–
Bangladesh	–	–	106	–	–	95	–	–	–	–
China	2,700	–	–	–	–	–	2,911	2,000	–	–
Egypt	–	–	–	–	–	–	–	–	–	120
India	2,880	120	291	–	–	–	402	1,890	–	–
Indonesia	–	–	212	–	–	945	251	–	–	–
Jordan	–	–	–	260	–	–	–	–	–	–
Kenya	–	74	–	–	–	–	60	–	–	–
Lao	–	–	–	–	–	–	60	–	–	–
Malaysia	1,000	–	33	–	–	–	66	–	–	–
Pakistan	–	–	–	–	–	–	1,450	–	–	–
Philippines	–	–	–	–	–	–	–	–	615	40
Sri Lanka	–	–	–	–	169	–	195	–	–	–
Syrian	–	–	–	660	–	700	–	–	–	–
Thailand	–	–	–	–	–	–	125	671	–	–
Uzbekistan	–	–	–	–	–	370	–	–	–	–
Vietnam	600	–	–	–	–	1,091	850	–	–	–

Source: JICA.

References

EIA (US Energy Information Administration) (2010) *International Energy Outlook 2010*, Washington, DC: US Department of Energy. Online, available at: www.eia.gov/oiaf/ieo.index.html (accessed May 20, 2011).

IEA (International Energy Agency) (2008a), *Energy Balances of OECD/Non-OECD Countries 2008*; CD-ROM, Paris: OECD.

IEA (International Energy Agency) (2008b) *CO2 Emissions From Fuel Combustion*; CD-ROM, Paris: OECD.

IEA (International Energy Agency) (2010) *World Energy Outlook 2010*, Paris: OECD. Online, available at: www.worldenergyoutlook.org/electricity.asp (accessed May 20, 2011).

JICA (n.d.) "Ex-Post Evaluation (ODA Loan)," JICA. Online, available at: www.jica.go.jp/english/operations/evaluation/index.html (accessed May 20, 2011).

World Bank (2010) *World Development Indicators 2010*; CD-ROM, Washington, DC: WB.

10

RURAL ELECTRIFICATION IN BANGLADESH

Implications for climate change mitigation

Shinji Kaneko, Satoru Komatsu, and Partha Pratim Ghosh

Introduction

Since the electric power sector is one of the major sources of greenhouse gas emissions in both the developed and developing worlds, the way electricity is generated is an important factor in climate change mitigation in most countries. Figure 10.1

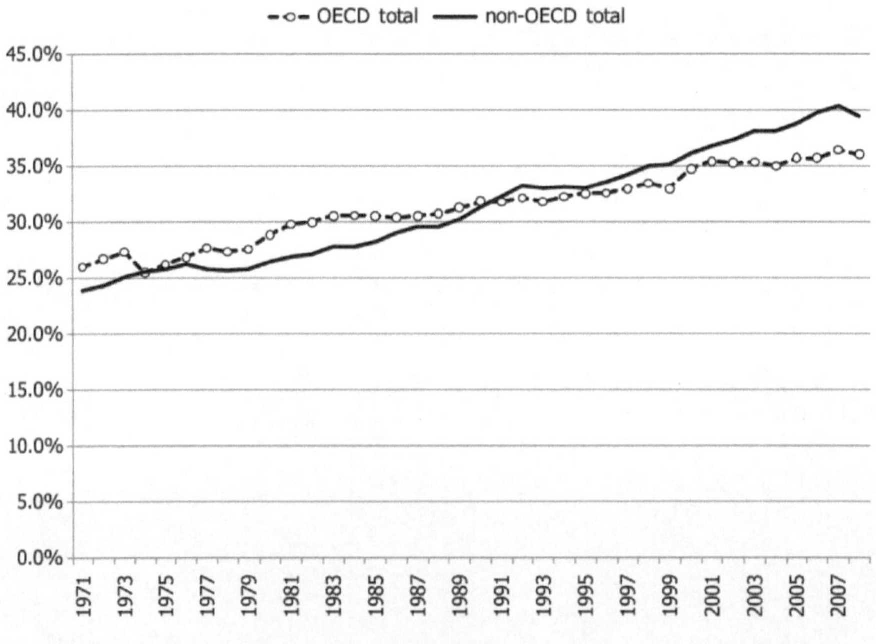

FIGURE 10.1 Share of energy related CO_2 emissions in electricity and heat sector to total emissions (source: IEA, 2010).

illustrates the historical trends of the contribution made by the power sector[1] to total emissions of energy related CO_2 in Organization for Economic Cooperation and Development (OECD) and non-OECD countries since 1971. The shares of the power sector in 1971 were approximately 25 percent in both groups. Subsequently, the figures have been similarly rising, to around 35 percent in 2000. A closer look at recent trends after 2000 reveals that the trends have decoupled, partly because of the rapid growth in China.

On the other hand, electrification is regarded as a priority area and as a means for national development and poverty alleviation policies in lesser developed countries. Significant capacity expansion of electricity generation is therefore anticipated for developing countries under future economic development scenarios. In 2008, the installed capacities of electricity generation in OECD countries and non-OECD countries were 881 GW and 2,173 GW, respectively, whereas respective total generation was 10,673 TWh and 9,510 TWh in the same year (IEA, 2010). The World Energy Outlook 2010 reported that under the Business as Usual (BaU) scenario,[2] these figures are projected to increase to 3,350 GW and 5,525 GW for installed capacity and 13,650 TWh and 24,773 TWh for electricity generation in OECD and non-OECD countries, respectively. In this context, future dependence on coal power generation is an important indicator for climate change mitigation. Whereas the capacity of coal power plants will be decreased in OECD countries by 2035 under the BaU scenario, that of non-OECD countries will be almost tripled, such that coal as a share of power generation will expand from 39.5 percent in 2008 to 43.9 percent in 2035 (Table 10.1). This implies that economically rational least-cost strategies under the current policy and regime would result in high carbon development in most developing countries as part of the process of their long-term economic development.

Amid concerns over the future prospect of conventional high carbon development patterns in the power sectors of developing countries, this chapter highlights a recent unique phenomenon in rural Bangladesh. Specifically, this is the stagnancy of grid rural electrification and success in off-grid rural electrification achieved by the rapid take-up of solar home systems (SHS) in areas where the likelihood of grid

TABLE 10.1 Electricity capacity and generation in 2008 and 2035 in OECD and non-OECD countries

	OECD			*Non-OECD*		
	2008	*2035*	*2035–2008*	*2008*	*2035*	*2035–2008*
Electricity capacity (GW)	881	1,219	338	2,173	5,525	3,352
Coal (GW)	198	162	−36	858	2,424	1,566
% of coal	22.5	13.3	−10.7	39.5	43.9	46.7
Electricity generation (TWh)	10,673	13,650	2,977	9,510	24,773	15,263
Coal (TWh)	3,882	4,012	130	4,391	12,443	8,052
% of coal	36.4	29.4	4.4	46.2	50.2	52.8

Source: IEA (2010).

connection is small in near future. We examine the phenomenon from the policy perspectives of climate change mitigation and international cooperation. The remainder of the chapter is organized as follows. The second section briefly introduces the past achievements and recent challenges of grid rural electrification in Bangladesh as a push factor behind the rapid application of solar technology in poor rural Bangladesh. The third section provides an overview on how and why a rapid increase in the SHS has taken place, followed by a brief summary of the consequences of installing SHS based on the results of our field survey. The fourth section compares the different policies and strategies of major donor agencies for the power sector development of the country. The final section discusses the challenges and opportunities inherent in the further expansion of SHS in rural Bangladesh and concludes the chapter with a proposal for a new international cooperation scheme with Japan.

Past achievements and current challenges of grid electrification in Bangladesh

History of rural electrification

When Bangladesh achieved independence in 1971, its electrification rate was only around 3 percent. The Constitution of Bangladesh was enacted in 1972 and Article 16, which covers rural development and agricultural reform, calls for rural development to be achieved through electrification. Rural electrification has consequently been a priority in national development plans from the earliest days of nationhood, and enormous efforts have been made.

Until 1977, the Bangladesh Power Development Board (BPDB) was the sole agency responsible for the generation, transmission, and distribution of electricity throughout the country. Because of operational difficulties, however, its activities were concentrated in urban (capital) areas and no government agency had responsibility for rural electrification at that time. To correct this situation, the Rural Electrification Board (REB) was established in 1977 to promote rural electrification as an autonomous government agency in support of the United States Agency for International Development (USAID).

The United States achieved electrification in its own rural areas by establishing consumer-owned electric cooperatives starting in the 1930s. It therefore has extensive experience in this area. The National Rural Electric Cooperative Association (NRECA) is the central organization of electric cooperatives in the United States. The international program division of the NRECA started development aid projects in 1962, dispatching more than 400 experts to 70 countries, mainly through a project on behalf of USAID. The rural electrification strategy in Bangladesh was designed by the NRECA adopting the cooperative model based on its lessons in the Philippines, such as the need for written instruction and a central agency, in addition to coping with severe corruption (USAID 2006). Therefore, the REB in Bangladesh was established as the central agency in Dhaka, independent from the

BPDB and other governmental organizations and with the major functions of providing supervision, technical support, and financial arrangements for each cooperative. The REB and the electric cooperatives established under the supervision of the REB have sole responsibility for electricity distribution, with electricity mostly generated by the BPDB and sold to the REB.

Anti-corruption system

Since the NRECA has carefully designed the initial institutional systems for rural electrification in Bangladesh, anti-corruption mechanisms are embedded and maintained in rural electrification projects in Bangladesh. The major business activities of the electric cooperatives, known locally as *Palli Bidyut Samity* (PBS) and established under the supervision of the REB, include: (i) planning and constructing grid connection infrastructure, such as electric cables and electric transformer stations; (ii) attracting new subscribers; and (iii) handling daily operations and customer service associated with electricity sales. The PBSs have an homogeneous organization, consisting of a board of directors (BoD), general manager (GM), employees, and customers. The BoD, which is responsible for important decisions on the operation and planning of PBS, consists of twelve to fifteen directors who are elected by customers every three years to give customers transparency and a sense of ownership. Politicians are not allowed to be directors, to avoid any possible political intervention in decision making. The GM, meanwhile, is a specialist and is responsible for management and daily operations of the PBS. The GM is appointed by the REB.

Each PBS does its own hiring, but employees are jointly trained at the REB and build their careers through programs that qualify them as electric engineers, accountants, managers, or other skilled personnel. In hiring PBS employees, special consideration is given to creating job opportunities for women in nearby communities. In addition, a PBS will have village and female advisors for intensive communications with villagers, especially to support women customers (refer to Figure 10.2).

One noteworthy instrument is the Performance Target Agreement (PTA), agreed annually between each PBS and the REB and setting out targets for twenty performance indicators, including the number of new subscribers, system loss, and the revenue collection rate. The REB provides economic incentives for all employees based on the principle of collective responsibility and depending on performance against the PTA, in the forms of bonuses or penalties. The PTA successfully motivates employees and GM to grow and improve performance, while also encouraging competition among the PBSs. Other components of the anti-corruption system can be summarized as follows (USAID 2006).

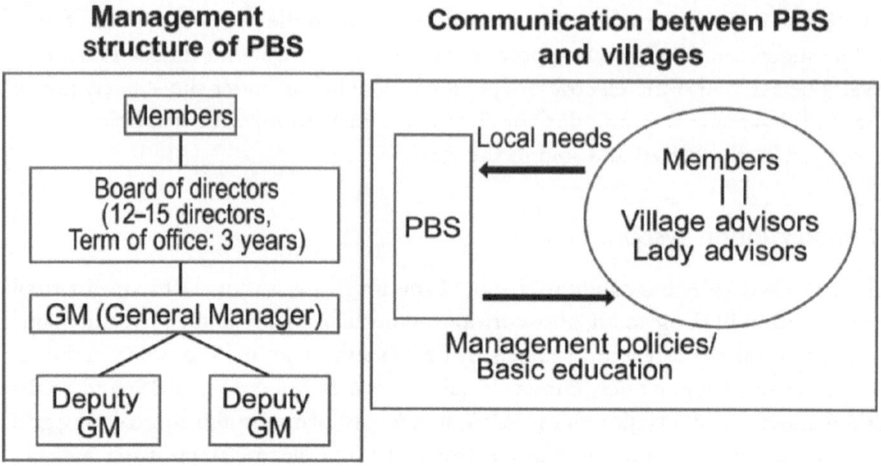

FIGURE 10.2 PBS organizational chart (source: authors).

Metering and bill collection

Electric consumption for each subscriber is measured entirely by separate meter. Meter readers are carefully selected and employed on an annual contract basis and not permitted to be hired by the same PBS for more than two years. The meters are changed every three months for a regular technical inspection. Electricity bills are paid at a bank or PBS office. Staff handling payment from each subscriber are rotated annually and supervised by the manager of the collection division.

Disconnecting clients

If a subscriber is more than two months late with payment, their electricity is disconnected. To be reconnected, the subscriber must pay a penalty of 10 percent of the amount in arrears.

Removing unauthorized connections

Regular patrols to detect illegal cables are undertaken in collaboration with customers and illegal connectors are penalized. In addition, if they turn out to have legal connection contracts to the grid, an additional penalty will apply.

Preventing the theft of distribution equipment

When a transformer or other equipment belonging to PBS are stolen or destroyed, users must pay some or all of its replacement cost.

Exclusion of political parties

As already mentioned, not only politicians but also those involved in partisan politics are ineligible for election as PBS officers. A strict application of economic, commercial, and technical criteria for determining priorities in grid expansion and thereby new connections are also highly effective in preventing political intervention.

Prohibition of unions

To prevent corruption, a cooperative law prohibits the forming of a labor union (although staff welfare organizations exist). This measure is adopted based on experience with the corruption of the BPDB and the Dhaka Electric Supply Authority (DESA), where labor unions defended corrupt employees from penalties.

Track record

Owing to the carefully and effectively designed institutions and highly motivated staff in the REB and PBSs, rural electrification has been producing a steady flow of successful results. Since the initial pilot project established thirteen PBSs by NRECA, the number of new PBSs has increased every year to seventy in total by 2005, covering most of the areas of the country. This success was also aided by financial support from multiple international donor agencies, attracted to an incorrupt sector in a least developed country with promising and tangible outcomes.

To show how well the REB has worked, some of the performance indicators of the REB are compared with other electricity distributors, namely BPDB, which covers urban areas across the country except for Dhaka city and its surrounding areas, as well as Dhaka Electric Supply Company Limited (DESCO) and DESA, which both cover Dhaka and its surrounding areas. The number of contract households has been growing rapidly under the REB, whereas those under other distributors remain low, as indicated in panel (A) in Figure 10.3. Meanwhile, panel (B) shows that electricity sales of the REB have been growing, but far below that of BPDB, which operates in urban areas other than the Dhaka capital area. One of the reasons for this is that sales include not only households but other types of contracts such as industry, which consume more electricity and are located in urban areas. Panels (C) and (D) compare two basic indicators measuring the quality of business operations of electricity distribution, system losses, and the fee collection rate. Both indicators are affected by the level of corruption. Although the system losses combining both technical and non-technical losses of electricity distributors have generally been improving in Bangladesh, REB is consistently the best performer (data were only available since 2001 in the panel (C)). The fee collection rate of REB is also high (nearly 100 percent) as well, shown in panel (D).

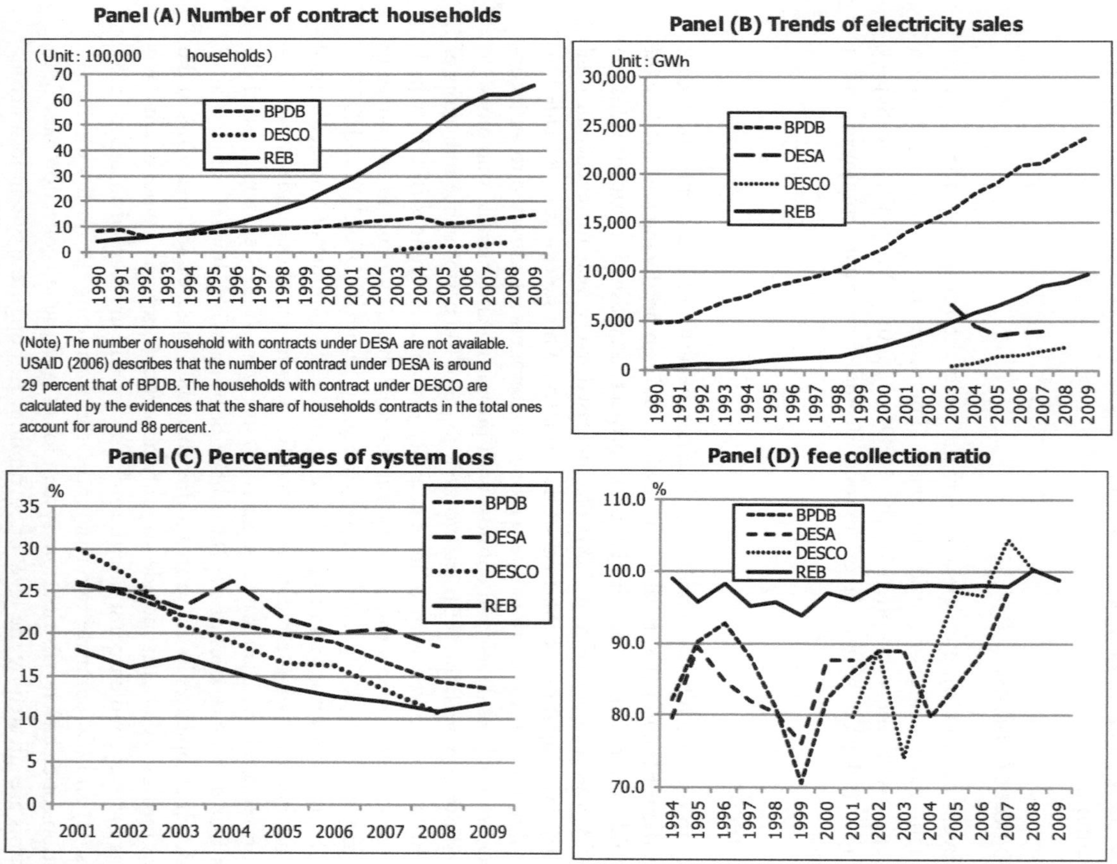

FIGURE 10.3 Comparison of performance indicators (source: authors modified from BPDB (2010), DESCO (2009), REB (2007, 2008, 2009)).

Current and future challenges

Despite the fact that rural electrification was generally successful and the number of electrified families has increased significantly, current per capita electricity consumption in Bangladesh is still among the lowest in the world, at just 183 kWh in 2009 (Koyama 2010). The household electrification rate in rural Bangladesh was also only 28 percent in 2008, effectively meaning that 94.9 million people live without electricity. As of June 2009, peak demand for electricity was 6,066 MW, while the supply capacity of electricity was only 4,162 MW, or approximately 70 percent of peak demand (Koyama 2010).

Chronic power outages adversely affect the business performance of PBSs and are more than occasional. A serious shortage of electricity troubled farmers during peak irrigation season in February 2006. The farmers eventually mobbed their local PBS, which resulted in seventeen fatalities among the farmers. There is a sad irony in this incident, as it was the diligent efforts by PBSs to successfully add customers in villages that raised expectations for electrification, which in turn extended the gap between demand and supply. It is clear that the root of the problem is exclusively on the supply side and the disconnect of governance (Bai *et al.* 2009: 255).

The ruling party in national government in Bangladesh has been changing every five years, when national elections take place and the long-term power development plans are consistently suspended and/or renewed by new party. In particular, in the period between 2001 and 2005, only one new power plant was constructed, adding merely 80 MW to existing capacity.

The rural electrification program of the REB itself also faces several challenges and problems. The project began in economically and commercially advantageous areas, with higher population densities and proximity to urban areas. With the expansion of the grid lines, penetration into remote, geographically difficult, and less populated areas makes new investment less economically viable. At the same time, the larger scale of business in the REB encourages politicians and bureaucrats in other ministries to intervene, challenging the autonomy of the REB. Our empirical study on the operating efficiency of PBSs confirms the significant negative relationship between the proportion of members of Parliament belonging to the ruling party and the operating efficiency in each PBS (Taniguchi and Kaneko 2009: 2433).

Why the rapid rise in SHS?

Background

The 'Vision Statement and Policy Statement on Power Sector Reforms' was adopted by the government of Bangladesh in 2000, with the goal of completing rural electrification by 2020. As noted, the household electrification rate is presently lagging far behind the goal, and the deadline will likely be missed. Given a chronic power shortage and an uncertain future, non-electrified households have begun to seek alternative sources of electricity. Since the early 2000s, the SHS has been booming in poor and remotely located villages, helped by non-government organizations such as Grameen Shakti and the BRAC Foundation.

The SHS is a package typically consisting of an imported photovoltaic (PV) panel producing several dozen watts, a domestically manufactured rechargeable battery, several 6 W fluorescent lamps, a mobile phone charger, a second-hand black and white TV, and a control panel. The price of a SHS ranges from approximately 300–1,000 USD depending on the size of the PV panel, which runs from 40–130 Wp, and other equipment. Rural households can purchase the SHS with either a single payment or under installment scheme. Since the average amount of electricity actually consumed is much smaller than that of grid electrification, the impact of electrification using SHSs are mostly characterized by an improvement in living standards without income generation effects. The following sections briefly explain the rapid growth in SHS with a focus on the key stakeholders.

Institutional perspective: role of IDCOL and NGOs

IDCOL

To understand the rapid expansion of SHS in Bangladesh, it is important to begin with the role being played by a government institution. A government-owned financial institution with a mandate of financing private and public sector firms in the infrastructure and renewable energy sectors, the Infrastructure Development Company Limited (IDCOL) started supporting the grassroots activities of non-governmental organizations (NGOs) installing SHS in January 2003. The support began with assistance from the International Development Association (IDA) of the World Bank and the Global Environment Facility (GEF). In addition to a stable and reliable supply of funds for NGO activities aiming to reduce the SHS price and build capacity in the forms of grants and soft loans, the IDCOL also plays a significant role in monitoring their activities as a complaint window for customers.

The initial installation target was set in 2003, calling for 50,000 SHS units to be installed in five-and-a-half years. This target was attained three years early, in August 2005. Other goals were also achieved ahead of schedule. The original plan for 200,000 installations by the end of 2009 proved to be very conservative, with 284,102 units installed by March 2009. As of December 2010, 763,957 SHS units had been installed under the IDCOL program (IDCOL 2011). Following this success, four donors, including the Kreditanstalt für Wiederaufbau (KfW), the German Technical Cooperation (GTZ), the Asian Development Bank (ADB), and the Islamic Development Bank, are supporting the IDCOL, in addition to the IDA and GEF.

Grameen Shakti

To date, thirty NGOs are working in rural Bangladesh under the program supported by IDCOL, selling SHS mostly through micro credit schemes. Among them, Grameen Shakti as one member organization of the Grameen Bank group is the largest contributor to the installation of SHS on the ground, accounting for 62.7 percent of all units installed under the IDCOL program as of December 2011 (IDCOL 2011). Grameen Shakti was established in June 1996 as a pioneer for private sector to become involved in solar electrification projects in Bangladesh, specializing in the promotion of renewable energy in Bangladesh (Barua *et al.* 2001). While implementing the program, Grameen Shakti strategically selected remote areas without conventional electricity supply and without any chance of being connected to the electricity grid within five to ten years (Barua 2001). Cumulative installation of SHS by Grameen Shakti has been accelerating over the last decade, reaching 317,591 units by 2009 (Grameen Shakti 2009) and 478,706 units by 2010 (IDCOL 2011). As a result, by 2010 Grameen Shakti had covered the country with 726 branch offices, 110 regional offices, twelve divisional offices, and forty-five technology centers, enabling it to provide customers with prompt and flexible service. It has also created many job opportunities and more than 3,000 female technicians and entrepreneurs have been trained. Several features of Grameen Shakti can be cited as factors for its success, including its dedicated, committed, and well trained staff, strict audit system, the strength of its micro credit and concessional payment scheme (down payments with a loan) to reduce the financial burden of the high up-front costs, quality control of products, a product guarantee service and after-sales support (providing equipment for maintenance and technical support), and identification of target consumers based on its extensive experience with micro credit programs (Kamal 2010).

Household perspectives: summary of survey results

Stated reasons by households

The following two sections summarize the major findings from our field survey in three villages in rural Bangladesh conducted in 2009. The survey covered not only SHS households but also non-SHS households as a control reference sample of the same size. The non-SHS do not have nearby SHS facilities in the same villages.

Table 10.2 shows the years of installation of SHS for the SHS respondents in the survey. The majority was installed after 2005. The range and distribution of size of the PV panels are presented in Table 10.3. Although larger packages with PV

TABLE 10.2 Installation of SHS by year (unit: number of households)

	2002	*2003*	*2004*	*2005*	*2006*	*2007*	*2008*	*Total*
Singair	0	1	1	14	10	36	40	102
Manikganj	0	0	2	0	24	30	45	101
Comilla	6	7	7	12	10	23	38	103

Source: authors.

TABLE 10.3 Contract packages of SHS (unit: number of households)

	40 Wp	*50 Wp*	*60 Wp*	*65 Wp*	*85 Wp*
Singair	16	39	3	32	9
Kishoreganj	32	53	0	13	3
Comilla	29	53	5	13	2

Source: authors.

Note
Four outliers in the sample respondents are found: 30 Wp, 75 Wp, 80 Wp in Singair and 130 Wp in Comilla, respectively.

panels producing more than 100 Wp are available, many respondents chose smaller packages.

Asked why they purchased an SHS, households gave the following reasons: (i) for a student studying at home; (ii) to watch TV; and (iii) to recharge the mobile phone at home. These are the major motivations for many households in the three villages, although the priority rankings differ slightly from village to village (Table 10.4). In non-electrified villages in rural Bangladesh, charging a mobile phone at a small market and/or at the house of a friend or relative is common practice. This is consistent with the fact that many households in Comilla village – the farthest from the nearest city – ranked charging a mobile phone as the number one priority.

Table 10.5 compares the number of mobile phones in both the SHS households and the non-SHS households. Of the 304 SHS households, 72.4 percent had at least one mobile phone before installing SHS, and 97.0 percent have at least one now. The average number of mobile phones has increased accordingly, from 0.93 to 1.60 units per household. Among SHS households that previously owned a mobile phone, 47.7 percent charged at the market and 22.3 percent charged at the house of a neighbor or friend before installing SHS.[3] On the other hand, 180 of the 308 non-SHS households (58.4 percent) currently have mobile phones, and the

TABLE 10.4 Common reasons and number of respondents (HH)

Village	*1st*	*2nd*	*3rd*	*4th*	*Total number of respondents*
Singair	Student study at home (27HH)	Watching TV (23HH)	Charging mobile phone (19HH)	Want electric lighting (19HH)	102 HH
Kishoreganj	Student study at home (41HH)	Want electric lighting (22HH)	Watching TV (22HH)	Charging mobile phone (16HH)	101 HH
Comilla	Charging mobile phone (16HH)	Student study at home (35HH)	Watching TV (15HH)	–	103 HH

Source: authors.

TABLE 10.5 Number of mobile phones

Number of mobile phones	*Number of non-SHS households*	*Number of SHS households*	
		Before	*Now*
0	128	84	9
1	157	172	166
2	16	37	91
3	6	7	21
4	1	4	10
5	0	0	7
Average number of mobile phones (per household)	0.69	0.93	1.60

Source: Komatsu *et al.* (2011a).

average number of mobile phones per household is 0.69. Of these 180 non-SHS households, 51.7 percent charge at the market and 41.1 percent charge at the house of a neighbor or friend.

Looking more closely at the practice of recharging mobile phones, we also compare the distance and time involved in traveling to recharging location and the frequency with which the trip is made. Table 10.6 shows the differences between the non-SHS households at present and the SHS households before installing the SHS in the past. The SHS households traveled longer distances taking a longer time, compared to that of non-SHS households. The frequency of recharging for the SHS households is also higher, even in the past. This implies that mobile phones could be one non-income factor in the decision to purchase the SHS package.

Determinants of SHS purchase decisions

With all data collected from the SHS households and non-SHS households from the three villages, we have additionally applied econometric techniques to investigate

TABLE 10.6 Distance and the frequency of visits to place for recharging mobile phones

	Non-SHS households	*SHS household (before)*
To market		
Distance (one way, km)	1.55	1.82
Distance (one way, minutes)	17.14	20.49
To neighbor's or friend's house		
Distance (one way, km)	0.10	0.32
Distance (one way, minutes)	6.03	7.82
Frequency of charging (per week, per each mobile)	2.76	3.21

Source: Komatsu *et al.* (2011a).

the determinants for two key decisions and improve our understanding the reasons (Komatsu *et al.* 2011b). One is the determinants for the decision to adopt SHS and the other is those for the choice of size of the panel, which represents the price of the total SHS package. The binary probit model and the ordered probit model are used for the former and latter cases, respectively. While household income is the primary factor in both decisions, we highlight non-income factors in the analyses, since household income is not by itself capable of fully explaining household adoption or the size of the SHS installed. As expected, after confirming that income is the primary determinant for both decisions, we found that non-income factors significantly improve the explanatory power of the models.

The binary probit model applied to the pooled data of both households with and without the SHS, reveals that ownership of rechargeable batteries, high kerosene consumption, and ownership of mobile phones are the key non-income determinants for adopting the SHS (Table 10.7). The ownership ratio of rechargeable batteries of SHS households declined markedly after the installation of SHS in three villages. As the batteries are usually very heavy to carry and involve transportation costs in addition to the direct cost of recharging, households with batteries had a strong incentive to adopt SHS. The reason for kerosene having a statistically significant influence on the adoption decision can be interpreted as being that the greater the kerosene consumption, the greater the incentive to save on the cost of kerosene.

TABLE 10.7 Determinants of decision for SHS adoption

	Model 4	*Model 5*	*Model 6*
Log_INC	0.461**	0.330*	0.362***
	($6.424*10^{-2}$)	($6.858*10^{-2}$)	($6.741*10^{-2}$)
BATTERY		0.788***	0.729***
		(0.119)	(0.119)
KERO		0.132***	
		($3.961*10^{-2}$)	
MOBILE			0.135*
			($7.956*10^{-2}$)
Cons.	–5.319***	–4.546***	–4.524***
	(0.740)	(0.116)	(0.763)
McFadden's R^2	0.067	0.132	0.123
Prob. $>\chi^2$	0.000	0.000	0.000

Source: Komatsu *et al.* (2011b).

Notes

Number of observations is 598 for each model. Values in parentheses are standard errors. *, *** denote significance at the 10%, 1% levels, respectively.

Log_INC: natural logarithm of total household income (taka, 2008 prices/year); BATTERY: whether rechargeable batteries are used (1 = Yes, 0 = No); KERO: kerosene consumption (liters/month); MOBILE: number of mobile phones.

While large differences in kerosene consumption between the SHS households before installation and the non-SHS households exist, many SHS households stopped using kerosene lamps after installing SHS. As for mobile phones, as explained before, ownership and use of a mobile phone is also confirmed as one of the key determinants for the decision to install SHS. The number of children and elderly persons were also important determinants.

The second analysis using the ordered probit model attempts to capture non-income as well as income factors affecting the choice of size (Table 10.8). The model applies to the sample of households with SHS. It identifies the factors as household income, kerosene consumption, number of children, concern about indoor pollution, and demand for electric lighting in the household. Among these factors, kerosene consumption shows relatively high elasticity in its influence on the decision to purchase larger SHS. As can be expected, household income is identified as one of the key determinants, but the model with income as the sole independent variable is again not capable of explaining the variation well. This finding is consistent with the sales practice of Grameen Shakti, where SHS size is

TABLE 10.8 Determinants of decision for SHS size selection

	Model 4	*Model 5*	*Model 6*
Log_INC	0.153**	0.135*	0.154**
	(6.937*10^{-2})	(7.329*10^{-2})	(7.417*10^{-2})
KERO		0.141***	0.144***
		(4.764*10^{-2})	(4.791*10^{-})
CHILD		7.836*10^{-2}*	9.075*10^{-2}**
		(4.279*10^{-2})	(4.311*10^{-2})
Pri_nopol			0.164***
			(6.304*10^{-2})
Pri_light			0.129***
			(3.607*10^{-2})
LOC_MAN		0.740***	0.677***
		(0.141)	(0.144)
McFadden's R^2	0.006	0.051	0.074
Prob. $>\chi^2$	0.027	0.000	0.000

Source: Komatsu *et al.* (2011b).

Notes

Number of observations is 296 for each model. Values in parentheses are standard errors. *, **, *** denote significance at the 10%, 5%, 1% levels, respectively.

Log_INC: natural logarithm of total household income (taka, 2008 prices/year); KERO: kerosene consumption (liters/month); CHILD: number of children (under 12 years old); Pri_nopol: kerosene lamp emits harmful pollutants as a motive for installation (Likert scale 0–5); Pri_light: need for electric lights as a motive for installation (Likert scale 0–5); LOC_MAN: location dummy for Manikganj (1 = Manikganj, 0 = others).

determined primarily by the number of lights needed by the household, which is directly related to the prior consumption of kerosene for lighting. The reason why households with more children are motivated to install a larger SHS is that the children can be expected to spend more time in the evening studying with electric lighting from a larger SHS. Awareness and concern of health risks associated with kerosene lamps also influences the decision to choose a larger PV panel, which is captured by cognitive measurement of concerns about pollution risk and the presence of elderly family members. In summary, the results from two empirical analyses suggest that non-income factors need to be carefully considered in promoting the SHS business.

Static payback analysis

Most households purchase their SHS on a monthly installment plan, with payments over either twenty-four or thirty-six months. In this section, we simply compare the monthly repayment of the SHS households with the reduction in energy costs, measured as the changes in expenditure due to the installation of the SHS (Komatsu *et al.* 2011b). Though a dynamic evaluation would be required for a complete financial viability assessment, this section is limited to an evaluation of the static condition to give a broad understanding of the cost and benefit structure due to limitations on complete data for all equipment and accessories. Therefore, only a partial sample of the households under the repayment period at the time of the survey is subject to the analysis.

Energy cost savings here denote the avoidance of past expenditures arising from kerosene consumption and recharging batteries before SHS installation. The time value of any travel involved is not included. For monthly payments after installation, the cost arising from repairs of package equipment is considered when it is not covered by warrantee. We found from our sample that in all cases the rechargeable batteries were the only items requiring repair. Table 10.9 summarizes the components for the comparison.

Figure 10.4 compares monthly expenditure on the SHS and energy cost savings for each major SHS package, based on the most common panel sizes. In each package, the left bar indicates the monthly payments for the SHS in addition to the expenditure on battery repair, while the right bar shows the energy cost savings. While the cost of repairs is very limited, the majority of energy cost savings comes from reduced kerosene consumption, followed by the avoided transportation costs associated with rechargeable batteries.

It is found that the transportation costs associated with rechargeable batteries are higher than the charging costs, which reflect the weight of the batteries. While repayments rise with the size of the panel, the energy cost saving is relatively equal across panels of different sizes. As the kerosene consumption of a large package holder was previously higher, the reduction in kerosene costs is larger. However, the difference in energy cost savings between a larger package and a smaller one is marginal, hence economic viability is higher for households with a smaller package.

TABLE 10.9 Comparison of monthly expenditure on SHS and fuel cost reduction

Energy cost saving		
Kerosene	Energy cost	Decrease in expenditure on kerosene
	Transportation cost	Decrease in the transportation cost for kerosene (fee for rickshaw cart, etc.)
Rechargeable batteries	Energy cost	Decrease in expenditure on charging rechargeable batteries[1]
	Transportation cost	Decrease in the transportation cost for rechargeable batteries (fee for rickshaw cart, etc.)
Monthly expenditure on SHS		
Monthly repayments for SHS		
Expenditure on the repair of batteries (calculated by the frequency of repair and the expenditure needed for each repair)[2]		

Source: Komatsu *et al.* (2011a).

Notes

1 We assume the benefits for those that continue to use rechargeable batteries after purchasing SHS are zero.

2 The SHS users also noted problems with the charge controllers, but these are frequently repaired at no charge. We have no information on any breakdowns in the solar panels.

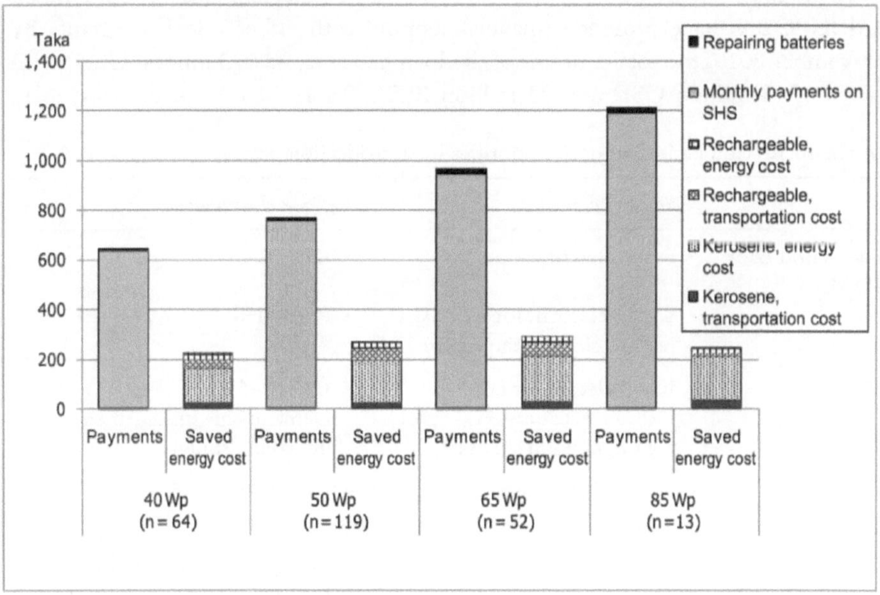

FIGURE 10.4 Comparison of monthly payments on SHS and energy savings (source: Kamatsu *et al.*, 2011a).

This analysis also supports the limited economic benefits and implies the significance of non-economic expectations and satisfaction from the SHS such as education and health impacts.

International support: programs of major donors

The current policies and programs of major donors are compared in this section. The government of Bangladesh relies heavily on international support from donors in many developmental sectors and this is true also of the power sector. Table 10.10 briefly summarizes key donors contributing to the power sector in Bangladesh. There are two main areas of international support: grid electrification and off-grid electrification. The World Bank and ADB support grid electrification both in rural areas and in cities, whereas JICA mostly supports grid rural electrification. In contrast, some donors such as USAID, DFID (Department for International Development, UK), and GTZ/KfW have supported grid rural electrification, but they appear to be scaling back their support.

Most donors also invest in off-grid electrification through IDCOL, with the exceptions of DFID, which supports grid rural electrification, and JICA, although JICA does provide indirect support to IDCOL through the ADB. Since the establishment of IDCOL in 1997, the World Bank (IDA), GTZ/KfW, ADB, Islamic Development Bank (IDB), and the Global Partnership on Output-Based Aid (GPOBA) have provided financial support with either a loan or grant. By November 2010, the total amount of the loan had reached 173 million USD, with grants of 14 million USD and 23 million EUR (Table 10.11). Of the total loan,

TABLE 10.10 Comparison of donor priorities in electrification projects

	Grid electrification	*Off-grid electrification*
The World Bank (IDA)	• Rural electrification	• IDCOL SHS
	• Urban electrification (Dhaka region, cities in the south)	• REB SHS
ADB	• Rural electrification • Urban electrification (Dhaka region, cities in the west)	• IDCOL SHS • Demand-side management (feasibility study)
USAID	• Rural electrification (scaling down)	• IDCOL SHS
DFID	• Rural electrification (scaling down)	• Pilot projects from multi donors found for climate change
GTZ/KfW	• Rural electrification (scaling down)	• IDCOL SHS • Renewable energy program
JICA	• Rural electrification	• Under considerations

Source: modified from presentation material by Koyama (2010).

TABLE 10.11 Record of support for IDCOL by each donor agency

	Loan	*Grant*
IDA and GEF	124 million USD	7 million USD
KfW	–	15 million EUR
GTZ	–	8 million EUR
ADB	33 million USD	–
IDB	16 million USD	–
GPOBA	–	7 million USD
Total	173 million USD	14 million USD and 23 million EUR

Source: IDCOL (2010).

Notes

IDA: International Development Agency; GEF: Global Environment Facility; GTZ: German Technical Cooperation; KfW: Kreditanstalt für Wiederaufbau; ADB: Asian Development Bank; IDB: Islamic Development Bank; GPOBA: Global Partnership on Output-Based Aid.

71.7 percent has been provided by the World Bank (IDA). In addition, the USAID directly supported Grameen Shakti with USD1.18 million during 2005 and 2008.

It is noted that GTZ/KfW emphasize technical support and demonstration experiments manifest two unique strategies: (i) targeting support for poorer households with small SHS (10–21 Wp) and pico SHS (3–6 Wp) exclusively for lighting and (ii) solar electrification is combined with other renewable energy uses as an integrated renewable energy program. GTZ/KfW installed 24,500 units of small SHS from 2007 to 2009 (Khaleq-uz-zaman 2010).

To the best of our knowledge, two areas of support are separately targeted by major donors, and considerable coordination and demarcation has been made in grid electrification and off-grid electrification through REB and IDCOL, respectively. However, there seems to be no clear consensus among the donors on how the two areas should be integrated or indeed whether they even should be.

Challenges to enlarge the scale of solar electrification

In general, the impacts of rural electrification on households can be categorized into two types, income generation effects and livelihoods improvement effects, although it may not be easy to clearly distinguish between the two over the longer term. The income generation effects of grid electrification in rural Bangladesh have been well documented and quantified (Barkat *et al.* 2002; Togawa and Kaneko 2006). For example, the lighting allows housewives to work at home into the evening, leading to higher income. In a similar vein, the elimination of time spent recharging a battery and/or mobile phone at the market also offers the possibility of additional income. Grid electricity enables farmers to introduce electric equipment for agriculture, which significantly improves productivity. In Bangladesh, electric pumping for irrigation from deep wells sometimes improves dramatically with a shift from single cropping to triple cropping of rice. For owners of small shops and restaurants, electricity offers more business opportunities such as the extension of business hours and installation of electric fans and fridges. Grid electrification also attracts industries and enterprises to establish new factories in rural areas, leading to job creation for villagers. These examples directly and indirectly generate incremental income sources, not only for electrified households but for all villagers.

On the other hand, the power generated by SHS and stored by rechargeable batteries is quite limited, and so the income generation effect in the shorter term is marginal. It can be roughly estimated to be less than one-fifth compared with the effect from average electricity consumption for electrified households with a grid connection under the current restricted supply.[4]

During our field survey, several parents and school teachers told us that children's performance in school improved remarkably after the installation of SHS. However, the income generation effects are not clearly captured as the major benefits of the SHS from our survey. Most of the tangible benefits from the SHS can therefore be generally understood as livelihood improvement effects, and not income generation effects. As a consequence, at this stage it is still only the wealthier families among the rural farmers – typically with remittance income – who can afford to purchase SHS, which is seen as a luxury commodity. Because of the insignificant income generation effect, poor families seem hesitant to purchase SHS, which will likely increase their debt. Figure 10.5 supports this view, comparing income prior to the installation of SHS between non-SHS households and SHS households.

Even the better-off families in rural Bangladesh, however, are still around the poverty line, and one might be surprised to learn that they can afford to purchase

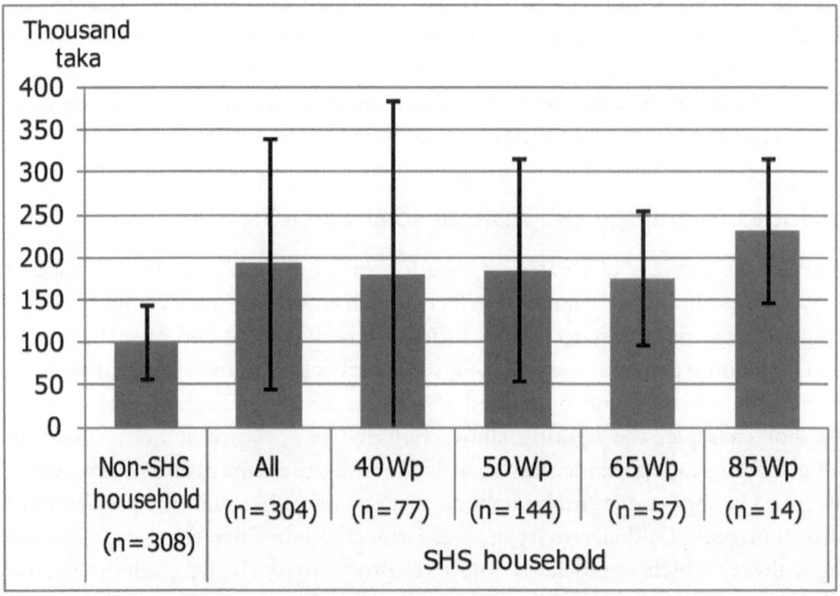

FIGURE 10.5 Incomes of non-SHS households and SHS households before installation (source: Komatsu *et al.*, 2011b).

one of the most expensive renewable sources of electricity. Production costs per kWh electricity with coal would be around 3 and 4 taka, while solar power would cost 17 taka. The cost from a solar power system comes down to 10 taka per kWh if an interest free loan is available, but it is still an expensive option (Hossain 2009). It would be economically more rational to promote grid electrification with cheaper energy resources like coal and natural gas. In addition, the total cumulative capacity of the SHS is still some 24 MW, while that of grid electrification is more than 4,000 MW. Therefore, even with the remarkable success of solar electrification, solar energy cannot meet total electricity demand. The economically rational scenario of power development in Bangladesh will place the rate of energy related CO_2 emissions to total emissions on an upward trajectory similar to that seen in China and India, as shown in Figure 10.6.

Is the micro demand for renewable energy with low carbon high technology in rural Bangladesh negligible? The price of a PV panel in Bangladesh is around 200 USD for 40 Wp, one of the most popular sizes, which represents for 5 USD per unit Wp. More than 500,000 panels are being sold in Bangladesh each year.

In Japan, a 4,000 Wp panel costs approximately 2.5 million yen, consisting of 1.7 million yen for the panel and 0.8 million yen for other costs (inverter/converter and installation). Assuming an exchange rate of 85 yen per USD, the unit cost of the panel in Japan is also 5 USD per Wp. Although the cost of solar panels has been slowly declining in recent years in Japan, subsidies are still needed to expand solar

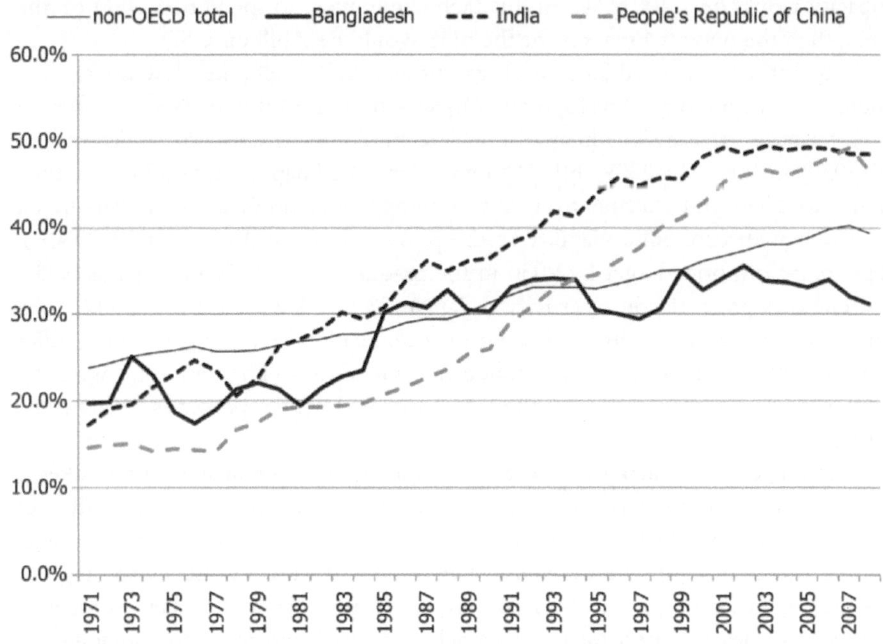

FIGURE 10.6 Ratio of energy related CO_2 emissions in electricity and heat sector to total emissions for Bangladesh, India, and China (source: IEA, 2010).

electrification in Japan. The subsidy is provided in the form of not only a capital subsidy but also as a feed-in tariff, which was introduced in 2010. Clearly, solar electrification is quite expensive even for some of the wealthiest consumers in the world, and the government needs to provide large economic incentives to stimulate demand. A significant breakthrough in cost savings is needed even for the market in Japan.

Returning to Bangladesh, one of the poorest countries in the world, it seems a puzzle that such expensive PV cells are rapidly penetrating non-electrified households with a subsistence lifestyle in rural areas. It does perhaps become more understandable from the perspective of the bottom/base of the pyramid (BOP) in the energy field. The BOP business breaks a product or service into micro pieces and creates a new market for the poor without discounting the unit price of the product or service. The World Resource Institute (2007) estimated the global BOP household market for energy at $228 billion annually for a global market of 2.1 billion people, in which Asia (including the Middle East) is the largest market as 177 billion USD, representing 1.5 billion people.[5]

There are approximately twenty million rural households in Bangladesh today, of which 5.5 million have electricity via a grid connection. Even if around 0.6 million households with the SHS are added to electrified households, some fourteen million households still remain without access to power. Assuming that the price of a PV cell drops by half, a 100 Wp SHS package would cost about 500 USD. If every non-electrified household were to install a 100 Wp SHS, the potential capacity would be 1,400 MW. And if each family were to spend 500 USD on the SHS, then the potential market for the SHS would be 7 billion USD.

In response to the need for a cost breakthrough in PV cells, the New Energy and Industrial Technology Development Organization (NEDO) in Japan adopted a long-term technology roadmap for PV technologies toward 2030 (known as PV2030) in 2004. In 2009, NEDO revised the roadmap by extending the time frame to 2050 and shifting from a technology orientation to a market-driven approach with economic viability as the primary priority (known as PV2030+). The long-term price target by 2030 in the roadmap is 7 yen/kWh, far below the current sales price of grid electricity in Japan (23 yen/kWh) (NEDO 2009). The target is expected to be attained using next-generation technologies which differ from the current mainstream technologies using silicon. The cost issue with the new technologies is to accelerate the R&D and deployment processes to attain the target by 2030 or even earlier.

The process of reducing the costs of alternative new products in the market is conceptualized and modeled by learning curve analysis and applies to the analysis of PV cells (i.e., Asano 2010). When a new technology becomes available in a market at an initial price of C(0), as illustrated in Figure 10.7, the price is usually too high to compete with existing products. There are two major technical challenges for new PV technologies that affect the price, namely conversion efficiency and durability.

If a government wants to accelerate the speed at which costs are reduced, it needs to provide financial support to compensate for the technological disadvantages and

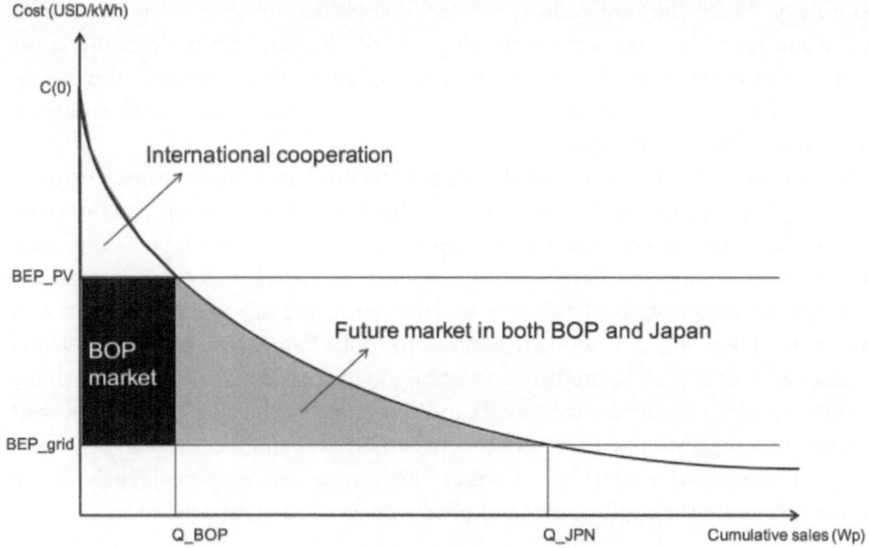

FIGURE 10.7 International cooperation and BOP in learning investment for new PV technology (source: authors).

stimulate demand. This applies with the goal of achieving market penetration for the next generation of PV cells. The market in Japan is already mature and thus a high conversion efficiency rate is equally required for new technologies to compete with existing technologies, due to severe space constraints. In contrast, the market of rural Bangladesh requires a much smaller panel, which can relax the space constraints. Thus, the less efficient new technology can be used as a component of the SHS package. If a sufficient warranty service is provided to compensate for the shorter lifetime through international cooperation with a well designed BOP business model, a new market for the second generation of PV technology could be created in rural Bangladesh.

The learning curve is then expected to decline with accumulation of stocks in the market (cumulative sales). The cost of power generation first reaches breakeven point (BEP_PV) at the same price as that of existing PV cells currently deployed in the market. The white area represents the learning investment necessary for new PV technology to be economically competitive with conventional technology, whereas the combined area of white and black areas represents the learning investment to become economically competitive with the grid electricity supply. Since PV technology currently available in Bangladesh is first generation, the black area can be regarded as part of the BOP market for SHS. If the white area is covered by international cooperation, the new technology should start penetrating the BOP market. Once the cumulative installation of new PV cells in the BOP market reaches Q_BOP with a price of less than BEP_PV, the market in Japan could be combined with the BOP market for the further deployment of the new technology. Since the cost reflects the technological features in conversion efficiency and durability, the new technology would become technologically competitive with existing technology by the point of BEP_PV. Moreover, the cost of the new PV technology would be lower than the BEP_PV, further accelerating the penetration of SHS in Bangladesh.

In contrast with the conventional pattern of technology transfer from developed to developing nations, where the type of technologies transferred is *proven technology*, the above idea attempts to transfer *appropriate technology*. One type of appropriate technology we propose here is still in its infancy, but is believed to be appropriate. Earlier commercialization of the new technology in the appropriate markets will bring mutual benefits, as a win–win strategy for both Bangladesh and Japan. While the strategy provides an immediate power supply in rural Bangladesh, it also eliminates the need for learning investments and accelerates technological advancement in Japan. As a suggestion, collaboration between NEDO and JICA would be effective in facilitating this dynamic. Further theoretical and empirical analyses are required to verify the applicability and effectiveness of the proposed model.

Notes

1 Due to data limitations, the power sector includes not only electric power supply but also heat supply. It does not include self-generation and other energy industries.

2 The scenario assumes that existing policies will not change markedly. This scenario is used as a baseline scenario for comparison with climate policy scenarios.

3 Others charge their mobile phones at home using rechargeable batteries.

4 On September 28, 2010, we held an expert seminar in Dhaka, Bangladesh, with the participation of major stakeholders in rural electrification, including the chairperson of REB, the CEO of IDCOL, the acting director of Grameen Shakti, and others. The number was confirmed during the seminar.

5 According to the World Resource Institute (2007), the procedure for estimating BOP markets is based on the following formula.

> Measured sector expenditure/total expenditure = estimated regional sector expenditure/total regional income, which is then solved for estimated regional sector expenditure. This amounts to assuming that the average ratio of sector expenditure to total expenditure as sampled in a measured group of countries is a good estimator for the same ratio in another group of countries in the region for which income but not standardized expenditure data are available. It also assumes that total household income equals total household expenditure, an equivalence already assumed in the methodology for assembling the income survey data.

References

Asano, K. (2010) 'Learning Curve Analysis of Photovoltaics in Japan,' CRIEPI (Center Research Institute of Electric Power Industry) Research Report, No. Y09020. Online, available at: http://criepi.denken.or.jp/jp/kenkikaku/report/detail/Y09020.html (accessed March 13, 2011).

Bai, X., Wieczorek, A.J., Kaneko, S., Lisson, S., and Contreras, A. (2009) 'Enabling Sustainability Transitions in Asia: The Importance of Vertical and Horizontal Linkages,' *Technological Forecasting and Social Change*, 76: 255–66.

Barkat, A., Khan, S., Rahman, M., Zaman, S., Poddar, A., Halim, S., Ratna, N., Majid, M., Maksud, A., Karim, A., and Islam, S. (2002) *Economic and Social Impact Evaluation Study of the Rural Electrification Program in Bangladesh*, Dhaka: Human Development Research Center.

Barua, D.C. (2001) 'Strategy for Promotions and Development of Renewable Technologies in Bangladesh: Experience from Grameen Shakti,' *Renewable Energy*, 205–10.

Barua, D.C., Urmee, T.P., Kumar, S. and Bhattacharya, S.C. (2001) 'A Photovoltaic Solar Home Perspectives System Dissemination Model' *Progress in Photovoltaics*, 9: 313–22.

BPDB (Bangladesh Power Development Board) (2010) 'Annual Report 2008–2009,' Dhaka: BPDB. Online, available at: www.bpdb.gov.bd/bpdb/attachments/075_Annual%20 Report-10.pdf (accessed July 13, 2011).

DESCO (Dhaka Electric Supply Company Limited) (2009) 'Annual Report 2008,' Dhaka: DESCO.

Grameen Shakti (2009) 'Yearwise Installation of SHS,' Dhaka: Grameen Shakti. Online, available at: www.gshakti.org/ (accessed August 6, 2010).

Hossain, I. (2009) The keynote speech of the roundtable entitled 'Energy Source for Power Generation,' on May 17 (organized by Energy and Power and GTZ-SED). Online, available at: http://phulbarinews.wordpress.com/2009/06/10/energy-source-for-power-generation/ (accessed January 27, 2011).

IDCOL (Infrastructure Development Company Ltd.) (2010) 'IDCOL Solar Energy Program Status Report (up to November 2010),' Dhaka: IDCOL.

IDCOL (Infrastructure Development Company Ltd.) (2011) 'Progress with SHS's Installation up to 31 December 2010,' Dhaka: IDCOL. Online, available at: www.idcol.org/prjshsm2004.php (accessed February 3, 2011).

IEA (International Energy Agency) (2010) 'IEA CO_2 Emissions from Fuel Combustion Statistics (database),' doi: 10.1787/data-00430-en (accessed January 27, 2011).

IMF (International Monetary Fund) (2009) 'World Economic Outlook Database,' October.

Kamal, A. (2010) 'Paving the Way for a Green and Sustainable Future: Pioneering an Integrated Market-Based Approach to bring Clean and Affordable Energy to the Rural People,' Dhaka: Grameen Shakti.

Khaleq-uz-zaman, M. (2010) 'Current Activities, Future Perspectives and Challenges of GTZ,' Proceedings of the second Expert Meeting on Rural Electrification and Solar Energy in Bangladesh, Dhaka, September, 28.

Komatsu, S., Kaneko, S., and Ghosh, P.P. (2011a) 'Are Micro-Benefits Negligible? The Implications of the Rapid Expansion of Solar Home Systems (SHS) in Rural Bangladesh for Sustainable Development,' *Energy Policy*, 39: 4022–31.

Komatsu, S., Kaneko, S., Shrestha, R.M., Ghosh, P.P. (2011b) 'Nonincome Factors Behind the Purchase Decisions of Solar Home Systems in Rural Bangladesh,' *Energy and Sustainable Development*, 15: 284–92.

Koyama, T. (2010) 'The Assistance for the Rural Electrification by JICA and other Donors and the Prospect for the Future Cooperation: The Case of Bangladesh,' Presentation materials of the Japan Society of International Cooperation Annual Conference, December 4 (in Japanese).

NEDO (New Energy and Industrial Technology Development Organization) (2009) 'Outline of the Roadmap PV2030+ (plus).' Online, available at: www.nedo.go.jp/content/100116421.pdf (accessed March 14, 2011).

REB (Rural Electrification Board) (2007) 'Annual Report REB 2006–2007,' Dhaka: REB.

REB (2008, 2009) 'Management Information System (MIS) Statement,' Dhaka: REB.

Rural Power for Poverty Reduction Program (Undated) 'Rural Electrification Board of Bangladesh and the Palli Bidyut Samities (PBSs).' Online, available at: http://pdf.usaid.gov/pdf_docs/PNACM053.pdf (accessed August 22, 2011).

Taniguchi, M. and Kaneko, S. (2009) 'Operational Performance of the Bangladesh Rural Electrification Program and its Determinants with a Focus on Political Interference,' *Energy Policy*, 37: 2433–39.

Togawa, M. and Kaneko, S. (2006) 'Area Coverage Rural Electrification Project (Phase IV-C), Ex-post Evaluation of ODA Loan Projects,' Tokyo: JICA. Online, available at: www.jica.go.jp/english/operations/evaluation/oda_loan/post/2006/pdf/project28_full.pdf (accessed March 14, 2011).

USAID (United States Agency for International Development) (2006) 'Integrity In Bangladesh's Rural Electrification,' USAID. Online, available at: http://pdf.usaid.gov/pdf_docs/PNADF759.pdf (accessed January 28, 2011).

World Resource Institute (2007) *The Next 4 Billion: Market Size and Business Strategy at the Base of the Pyramid*, Washington, DC: World Resource Institute. Online, available at: www.wri.org/publication/the-next-4-billion (accessed March 14, 2011).

11

ASIAN LOW-CARBON SOCIETY SCENARIO DEVELOPMENT AND POLICY IMPLEMENTATION

Junichi Fujino

Introduction

The development of mid- and long-term scenarios for the Asian region which show the ways to achieve a low carbon society (LCS) together with economic development, and with evaluation measures to realize these scenarios, is an issue of pressing concern. If we set the 2 degree target as the climate stabilization target, it is suggested that we need to cut greenhouse gas (GHG) emissions by half by 2050. We need to identify possible leapfrog pathways rather than conventional ones, using scientific findings.

AIM (Asian Pacific Integrated Model) was developed in 1990 for assessing policy options for stabilizing the global climate, particularly in the Asian-Pacific region, with the objectives of reducing GHG emissions and mitigating the impacts of climate change (AIM, 2011). AIM is a large-scale computer simulation model developed by the National Institute for Environmental Studies in collaboration with Kyoto University and several research institutes in the Asian-Pacific region. We have also research collaboration with Asian researchers from countries such as China, India, Korea, Thailand, Malaysia, Indonesia, Vietnam, Cambodia, and Bangladesh (NIES, 2011).

In this chapter, I would like to investigate the relationship between scientific findings – especially those based on simulation models – and policy implementation at the national level (Japan) and local level (Shiga prefecture and Kyoto City in Japan, and Iskandar in Malaysia).

Interaction between science and policy at the national level – Japan LCS scenarios

Japan has had on a few occasions the opportunity to make decisions concerning GHG emissions targets at the international level.

At COP3 in 1997, reductions were targeted at 6 percent, and at COP15 in 2010 a 25 percent reduction was targeted by 2020 and 80 percent by 2050.

While the author was not involved in the process and is not aware of the details, in 1997 both the Ministry for Economy, Trade and Industry and the Ministry of the Environment (MOE) carried out simulations based on their respective standpoints. In the end, the government finally agreed on a reduction target of just around 6 percent, with two stages for 2010.

The first was the Ministry of the Environment-sponsored "Japan Low-Carbon Society Scenario Research" which began in 2004 with the aim to analyze reductions until 2050.

The other was an analysis which began in 2009 to run until 2020 carried out in the chief cabinet – this being a comparative analysis of nationwide simulations in a mid-term plan implementation committee. I wish to analyze the role of the science and policy surrounding the targets adopted at COP15.

Involvement of Japanese LCS scenarios for 2050 in research and policy

The Japanese Low-Carbon Society Scenario Research with a five-year mid-term sponsorship of the MOE employs 60 researchers, comprising five teams (scenarios, system setting, city planning, ICT, and transport) and began to draw a picture as to how Japan could achieve a 60–80 percent reduction in GHG emissions by 2050.

The "2050 Japan Low-Carbon Society" scenario team delivered the report "Japan Scenarios towards a Low-Carbon Society (LCS) – a Feasibility study for 70% CO_2 emissions reduction by 2050 below the 1990 level" on February 15, 2007, and "A Dozen Actions towards Low-Carbon Societies (LCSs)" on May 24, 2008 ("2050 Japan Low-Carbon Society" scenario team 2008). It could be one of the scientific findings that made Former Prime Minister Abe launch his invitation to "Cool Earth 50 – 3 Proposals, 3 Principles" on May 24, 2007, aiming to set a long-term target of cutting global emissions by half from the current level by 2050 as a common goal for the entire world, and Former Prime Minister Fukuda to introduce the "Fukuda Vision" on June 9, 2008, stating that "Low-Carbon Society" is the key direction to have sustainable development in Japan, and set a target of 60–80 percent domestic GHG emissions reduction by 2050 from the current level. In addition, the cabinet approved the "Action plan to build a Low-Carbon Society" on July 29, 2008, and the policy was accepted at the government level in relation to the proposal initiative which concretely defined "A Dozen Actions towards LCSs." Following this, in August 2009, "Japan Roadmaps towards LCSs" was published, based on the results obtained from analyses utilizing a back-cast model. These analyses showed quantitatively the methodologies and necessary expenses needed for a 70 percent reduction by 2050 ("2050 Japan Low-Carbon Society" scenario team 2009).

With this research the following were made possible: (1) testing the feasibility of the vision of a 70 percent reduction by 2050; (2) proposing means for the realization of "A Dozen Actions towards a LCS"; and (3) testing the policy in relation to the existence of a quantifiable route using the back-cast model. The report also stated that based on these, the policy makers could determine a GHG emissions reduction standard both in Japan and worldwide, and that once the concrete direction of initiatives has been decided, the "low-carbon society action plan" could also be realized.

The role of science and policy in setting the GHG emissions reduction target by 2020

The process by which Japan arrived at its decision at COP3 in 1997 to reduce GHG emissions by 6 percent by 2020, as well as the role of the Ministry of Economy, Trade and Industry, the Ministry of the Environment and various expert groups supporting them have been regarded as lacking transparency.

With the COP15 in December 2010 in sight, those persons responsible for the development and management of the energy and economic model for Japan took part in the planning process toward Japan's decision in regards to GHG emissions reduction by 2020, with the aim to produce a technically based model for Japan as well as for the entire planet.

An interim review panel was established in November 2009, and at their seventh meeting on April 14, 2010, the panel indicated "Six Options" as targets for Japan's domestic efforts at reductions.

With the background of public comments and the results of opinion polls, Former Prime Minister Aso selected one option from this Options list and added 1 percent to the 2020 reduction target of 14 percent (compared to 2005) to make 15 percent (8 percent compared to 1990). This was made public at that year's SBSTA (Subsidiary Body for Scientific and Technological Advice) in June 2010.

When the government changed from Liberal Democrat Party (LDP) to Democratic Party of Japan (DPJ) at the end of August of the same year, Former Prime Minister Hatoyama made a declaration in his manifesto on the international stage in September 2010 about an international commitment for a 25 percent reduction by 2020. Moreover, this 25 percent reduction would not be just a national reduction measure that Aso's target focused on, but it would include Kyoto credits, and presupposed that all the important countries concerned would make comparable efforts.

With this background, a task force to focus on the issue of how a 25 percent reduction could be achieved by 2020 was convened on October 23.

The model researchers involved in the preparation of the "Six Options" now stepped up on the commission stage and were newly appointed as experts to further analyze and evaluate these results. However, they were not able to show the expected positive economic effects resulting from innovations which would come from the drastic emissions reduction, and it was therefore decided that November 19 would be the fifth and last time the group met at this moment.

There is a research report analyzing the role of the specialized knowledge of the decision-making process (Kubo 2011) according to which the role of science in the "Six Options" process was only an "information source" from which policy makers chose one option, while, in the "Task force" process, goals were decided in advance by the policy side and science was employed to help find a solution to achieve those goals.

This solution was to include constituents from the climate change bill as decreed by the MOEJ (Ministry of the Environment, Japan) and to be carried out on an ongoing basis under their auspices: a GHG reductions target, carbon tax, emission trading, Feed-In-Tariff (FIT) for renewable energy, and so on.

The "Mid- and Long-Term Roadmap Investigative Committee" met from December 2009 to March 2010, and from April 2010 a "Global Environmental Mid-to-Long-term Roadmap Subcommittee" was implemented.

Here, the knowledge of experts on the simulation model associated with the aforementioned decision-making process as well as concrete mitigation methods were evident and investigations proceeded in the form of working groups comprising experts including the areas of housing, buildings, transport and automobiles, urban structure, manufacturing, agriculture/forestry/fishery communities, and energy supply, in pursuit of the means to make emissions reduction a reality.

The hearings from involved parties, including over 100 experts in a variety of fields in a total of 60 working groups, and the accumulated knowledge of 19 subcommittee meetings were summarized in a report exceeding 1,500 pages in length. This was submitted to the ninety-second meeting of the Central Environmental Committee, Global Environmental Subcommittee on December 28, 2010 (MOEJ 2011).

The model simulation provided a quantitative dataset which enhanced discussions among experts from different fields, and those feedbacks improved the quality of the simulation results. With regard to information that could not be dealt with by model simulation due to the difference in the required geographical coverage such as country level and regional level, the reality level of the constructed scenarios could be increased by investigating the feasibility of the visions for each sector. The involvement of government administration (mainly from MOEJ) could enhance the possibility of policy implementation.

On the other hand it is hard to evaluate the objectiveness and validity of countermeasures and policy to achieve a challenging goal that no one has ever reached, such as a 25 percent GHG reduction by 2020 and 80 percent by 2050. It is therefore important to keep doing PDCA (plan, do, check, action) toward achieving ambitious targets, to monitor the climate policy impact and to analyze the relationship between cause and effect. It is quite interesting how science and simulation models can be effective for this PDCA cycle.

In fact, we did not have a climate change bill in Japan until September, 2011, due to 3/11 and chaos in domestic policy even though there was a chance for a bill

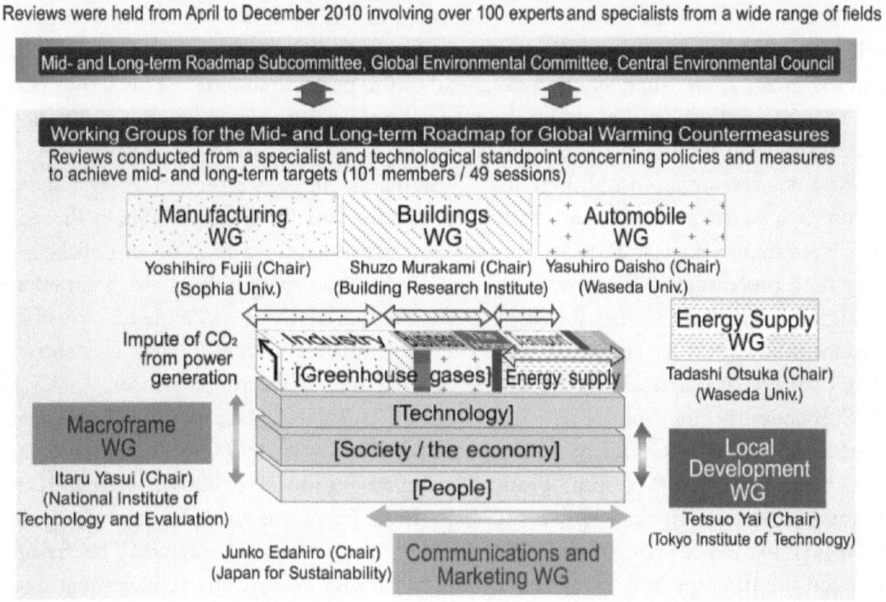

FIGURE 11.1 Multiple perspectives are necessary to consider mid- and long-term mitigation roadmaps (source: author).

to pass in summer 2010. We need to reconsider why we need low carbon societies and how we can achieve them (Fujino 2010).

Local-level interaction between science and policy – Shiga and Kyoto

Several members from Kyoto University and Ritsumeikan University, who joined the "Japan Low-Carbon Society Scenario Research" project during FY2004 and FY2008, supported the development of regional-level LCS scenarios such as those for Shiga prefecture and Kyoto City. These simulation results had a certain influence on setting up a regional environmental policy (mainly climate policy) and contributed to providing background information for local ordinance.

Shiga LCS scenarios

Shiga is one of the local governments which is most proactive in promoting sustainable society development. Shiga prefecture initiated internal discussions on how to develop a sustainable society in 2004. Lake Biwa is the symbol of Shiga prefecture because it is the largest lake in Japan and a mirror of the environmental situation in that both industrial and domestic pollutants affect it. Prefectural officers investigated several sustainable society indicators and finally selected CO_2 emissions as being one of the most important indicators.

Table 11.1 shows the history of development of visions and roadmaps for a sustainable society in Shiga. It is interesting to note that research groups always release visions and roadmaps approximately one year ahead of the prefectural office. This shows that visions and roadmaps produced by the researchers are very important for local policy making and the process to review and redevelop them takes around one year.

At first, researchers published the "Sustainable Shiga Vision" in 2006. Then, "Shiga's Scenario towards the Realization of a Sustainable Society" (Shiga Prefecture Sustainable Society Research Team 2007) was released in 2007 in collaboration with prefectural officers who were in charge of climate policy. Shiga aimed at reducing GHG emissions by half by 2030 compared to the 1990 level. Possible countermeasures were proposed based on the simulation model. Figure 11.2 shows that a variety of undertakings are necessary to achieve a drastic reduction.

Subsequently, the Shiga Prefecture Sustainable Society Research Team released "A Roadmap for Sustainable Shiga toward 2030" in October 2009 (Shiga Prefecture Sustainable Society Research Team: Roadmap Committee 2009), prepared by researchers at Ritsumeikan University, Lake Biwa Environmental Research Institute (LBERI) and Kyoto University, with advice from prefectural officers. The report showed the timeline and the necessary policy package toward the achievement of a

TABLE 11.1 Stakeholders' actions to develop sustainable society scenarios and roadmaps

Time	*Events*	*Stakeholders*
2004	Beginning of discussions in Shiga prefecture	Prefectural office
May 2005	Investigative Committee on Shiga Sustainable Society Vision established	Research group and Prefectural office
June 2005	"Lake Biwa Environmental Research Institute" (LBERI) established	Research group and Prefectural office
February 2006	Release of "Sustainable Shiga Vision"	Research group
April 2006	Beginning of investigations about sustainable society policy by former prefectural governor	Prefectural office
June 2006	New prefectural governor elected	Prefectural office
March 2007	Release of "Shiga's Scenario towards the Realization of a Sustainable Society"	Research group
March 2008	Release of "A Vision for Sustainable Shiga"	Prefectural office
October 2009	Release of "A Roadmap for Sustainable Shiga toward 2030"	Research group
August–October 2009	Opinion exchange among stakeholders in Shiga prefecture	Prefectural office
October 2010	Proposal: "Ordinance for the Promotion of Shiga Low-Carbon Society"	Prefectural office

Source: revised from Suda and Fujino (2010).

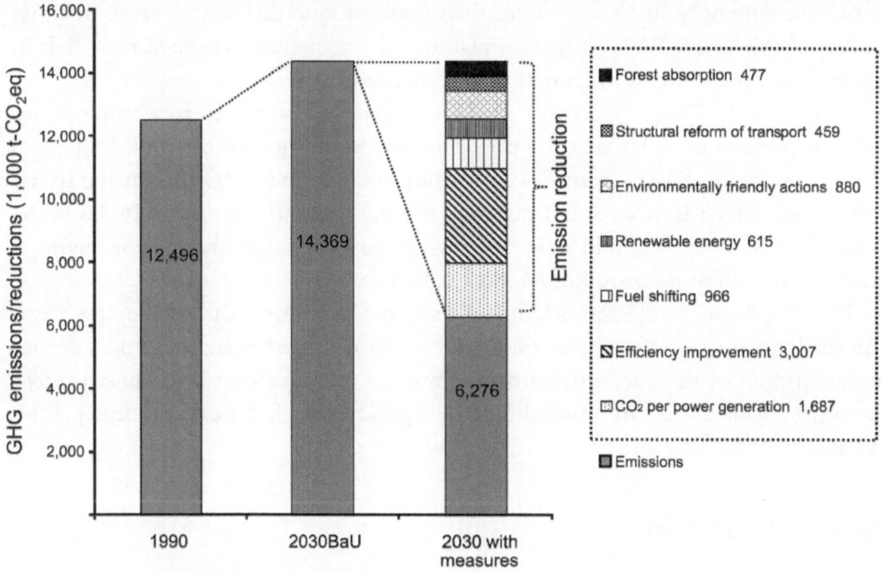

FIGURE 11.2 Contribution by mitigation measures in the Shiga scenario (source: Shiga Prefecture Sustainable Society Research Team 2007).

viable Shiga sustainable society vision. The other specific feature of this report is that it includes a calculation of the necessary investment (cost) to implement each countermeasure. Thereafter the Environmental Council of Shiga Prefecture released "To achieve a Low-Carbon Society in Shiga Prefecture" in March 2010. Based on local dialogues with citizens using information from available reports, Shiga Prefecture published the latest report "Roadmap toward Shiga Low-Carbon Society" in January 2011 (Shiga Prefecture 2011). This report also indicates the necessary public expenditure to achieve LCS in Shiga prefecture which is similar to the report in October 2009. The ordinance of promotion to develop Shiga LCS has been enforced since its approval at the regular meeting of Shiga prefectural assembly in February 2011. This ordinance prescribes the responsibility of all the stakeholders who are related to GHG emissions and absorptions in business activity, building environment, city planning, automobiles, conservation of forestry, agriculture and fishery, and so on.

Academic activity supporting administrative decisions is an interesting process. The researchers developed (1) 2030 visions and (2) roadmaps with a feasible combination of countermeasures and policies including expenditure, and approximately one year later, the prefecture released its own report based on discussions provoked by the researchers. If the prefecture side could understand the reactions of stakeholders on academic outputs, it could develop more acceptable visions and roadmaps for the stakeholders. This is the reason why there is approximately a one-year gap between the release of reports by the researchers and the prefecture. Specifically the role of "Lake Biwa Environmental Research Institute" (LBERI) is quite interesting. It has two faces, one as a local research institute and another as a local government sponsored institute. It is quite unique that this research institute published the latest GHG mitigation plan based on scientific simulation models in advance and contributed to climate policy making.

Originally the Shiga sustainable society concept was launched by a study group in Shiga prefecture. After that the previous and present governors took leadership to set the target to a 50 percent GHG emissions reduction by 2030 compared to the 1990 level, which is more ambitious than the current national target. It should be noted that the citizens and local businesses are very conscious about their environment because they are proud of Mother Lake Biwa.

The "Ordinance of Promotion to Develop Shiga Low-Carbon Society" was enforced in February 2011 based on these conditions. Further analyses and a deeper understanding of the background of this process are necessary, and Shiga prefecture's firm position and ambition should be a good example for other municipalities as well.

Kyoto LCS scenarios

Kyoto City is famous worldwide not just for receiving over 50 million tourists per year but also as the place which gave birth to the Kyoto Protocol. As part of the dissemination of this pioneering initiative, Kyoto City enacted the "Kyoto City Global Warming Policy" which would become the first law in Japan dedicated to tackling the issue of global warming.

This legislation put forward reduction targets aimed at exceeding 10 percent reduction of GHGs produced by cities by 2010 (compared to the 1990 level) with a nationwide target of a 6 percent reduction (0.6 percent at energy source). One further incentive for the acceleration of Kyoto City's global warming measures may have its origin in the initiative for the application and designation of Eco-model Cities in Japan. Eco-model Cities are cities which are exemplary on a world scale of furthering the transition to "Low-Carbon Societies." A major strategic initiative to this end was the administrative policy speech by Prime Minister Fukuda on January 18, 2008, which resulted in the "City and Lifestyles Development Plan" (January 29, 2008, approved by the Regional Regeneration Headquarters).

There were a total of 82 applications (89 groups) during the application period between April 11 and May 21, 2008, from Hokkaido to Okinawa and from diverse cities and regions, including both designated cities and small towns with a population of around 2,000 people. Many of the proposals set targets for wide-scale reductions in GHGs in the mid to long term (cut by half by 2050, 20–30 percent reduction in the medium term) and passionately argued for ways toward the realization of these goals – for example, measures which would simultaneously mitigate environmental burdens and promote sustainable development for the region while taking advantage of that region's particular characteristics.

Originally it was planned that ten local authorities would be selected, but on July 23, 2008, along with the designation of six local authorities as Eco-model Cities, it was also announced that an additional seven local authorities had been selected as

Reserve Eco-model Cities. Kyoto City was amongst these reserve cities. Subsequently on January 22, 2009, Kyoto, along with all the other local authorities which had been selected as reserve Eco-model Cities, was formally designated as an Eco-model City.

In this context, in April 2009, the Kyoto City Office completely rearranged its administrative structure, such that the Environmental Policy Office was positioned at the head of their administrative structure thus making it clear that the environment was at the very crux of city policy. After this, and in order to carry out a more concrete simulative analysis, with the cooperation of research groups from Kyoto and Ritsumeikan Universities they laid out the policy structure for scenarios which would result in a 25 percent reduction in GHG emissions by 2020 (compared to 1990), and a 40 percent reduction by 2030.

On the basis of the six actions described in the Kyoto Eco-model City Action Plan (Figure 11.3), a policy roadmap with the effects of those reductions and a timeline were published in August 2009. Based on the simulation results in Kyoto City, a modified version of the Kyoto City Global Warming Policy Document which contained the ambitious targets of a 25 percent reduction by 2020 and a 40 percent reduction by 2030 (compared to the 1990 level) was submitted to the city council and was approved unanimously in September 2010.

Now, what would Kyoto as a low carbon city be like? And who would have to do what, to achieve this goal, and when would they have to do it? The achievement of a low carbon city would require a citywide debate and action. Moreover, a great deal of information would be needed with regard to societal development and low carbon technologies.

The Research Team of Sustainable Society Kyoto has constructed a "Roadmap to a low-carbon society" for use as a basis for this discussion (Research Team of Sustainable Society Kyoto 2009). To forecast a future society scenario, first they assumed moderate economic growth of about 1.3 percent per year and a gradual decrease in population. As a point of reference, they used the most recent low carbon measure, entitled "Kyoto City Environmental Model City Action Plan," which was completed in March, 2009. The major findings are as follows.

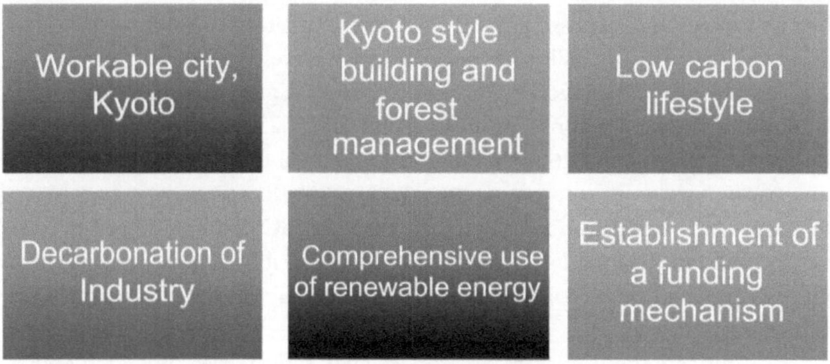

FIGURE 11.3 Six actions in environmental modeling City Kyoto plan (source: Research Team of Sustainable Society Kyoto 2009).

1. It is technically possible to reduce CO_2 emissions by 40 percent or more by 2030 compared to 1990. This will require the widespread adoption of actions far greater than the current situation in the usage of highly energy efficient equipment, better insulated buildings, the use of renewable energy, energy saving behavior, and other measures (Figure 11.3).
2. A path exists which will enable the necessary measures to be widely adopted in the 20 years between 2010 and 2030. If actions are promoted smoothly by 2025, 90 percent or more of the final measures can be implemented.
3. In order for measures to be disseminated, incentives must be provided as Kyoto Municipal Government policies. Most of the incentives are already included in the measures currently being planned by Kyoto City.
4. Most of the entities that must take action in order to achieve actual reductions are citizens and companies in the private sector. In line with this, the understanding and cooperation of the general public and companies in the private sector will be needed in order for the policies introduced by Kyoto City to be accepted and to lead to action.

Iskandar (Malaysia) scenarios

A research group in UTM (Universiti Technologie Malaysia), Kyoto University, Okayama University, and Ritsumeikan University released "Low-Carbon City 2025 – Sustainable Iskandar Malaysia," in January 2009. This report aimed to show the possibility of developing Iskandar Malaysia (IM) into a Low Carbon Region. The methodology involves (i) developing a current inventory of IM's GHG emissions, and (ii) quantifying the socioeconomic activity level in 2025 according to the IM Comprehensive Development Plan (CDP) 2025. According to the proposed development, the increase in GHG emissions is estimated based on the assumption of (1) 2025 BaU (Business as Usual – without mitigation measures) and (2) 2025 CM (with Counter Mitigation measures) as well as the potential to reduce the GHG emissions by low carbon measures available by 2025. The main findings are as follows.

1. Current annual GHG emissions of IM are approximately 12.6 million t-CO_2.
2. Under the scenario of "Without mitigation measures," GHG emissions will likely increase to 45.5 million t-CO_2, or 3.6 times higher than in 2005.
3. By adopting the mitigation options available by 2025, the emissions can be reduced by approximately 60 percent and suppressed to 19.6 million t-CO_2.

The current per capita emission for the IM area is 9.3 t-CO_2 (2005). This is greater than the national average of Malaysia (5.0 t). In 2025 it will likely increase to 15.1 t-CO_2/year without countermeasures and to 6.5 t-CO_2/year with countermeasures.

In order to mitigate the emissions to a lower level, several measures such as diffusion of low carbon technologies, energy efficient buildings and increased use of public transport and renewable energy should be introduced (Figure 11.4).

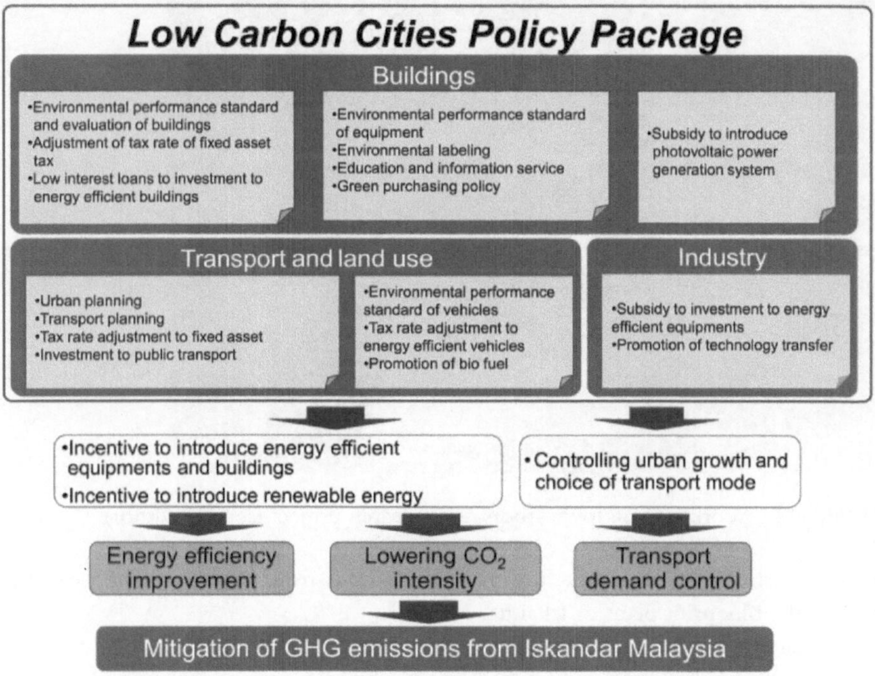

FIGURE 11.4 Low carbon cities policy packages in Iskandar Malaysia (source: UTMI *et al.* 2009).

Future steps

Recently we have coordinated a two-day symposium and workshop on "Low-carbon Asia Research Project" on July 4–5, 2011, in Johor Bahru (Figure 11.5). This was jointly organized by Universiti Teknologi Malaysia (UTM), Iskandar Regional Development Authority (IRDA), and Japanese counterparts, namely Kyoto University, the National Institute for Environmental Studies (NIES), and Japan International Cooperation Agency – Japan Science Technology Fund (JST), and JICA in conjunction with the launching of Iskandar Malaysia Low-Carbon Study to be officiated by Chief Minister of Johor State.

This symposium marked the beginning of capacity-building activities of the Science and Technology Research on Environment Projects (SATREPS) on Development of Low-Carbon Society Scenarios for the Asia Region in Malaysia. At the workshop, researchers from Asian regions, namely Indonesia, Thailand, China, and Cambodia, made presentations using AIM models in their respective countries.

It is important for developing countries such as Malaysia and other ASEAN countries to have capacity building in line with the UNFCCC framework through their own research. In this way, the policies for low carbon development can be promoted voluntarily by researchers of the respective country. The findings from this research can be implemented continuously in actual policy management.

FIGURE 11.5 Workshop on low carbon Asia research project (source: author).

This workshop will highlight how research output from AIM modeling can be used in the blueprint plans of Iskandar Malaysia.

Ultimately, this LCS research is to help initiate and facilitate the development of a research network in Asia on LCS and capacity building for researchers and officers at central and local governmental agencies, as well as provide opportunities and a platform for collaboration between policy makers and researchers.

When I looked at the decision-making process to develop scenarios and policy roadmaps toward LCS in Japan, Shiga prefecture, Kyoto City, I realized that good communication between experts and administrative offices was essential to have a solid plan. In December 2010, the Ministry of the Environment developed mid- and long-term GHG mitigation roadmaps. Shiga prefecture has enforced an ordinance that links with budgeting for GHG mitigation policy, and Kyoto City has developed an ordinance that includes ambitious GHG reduction targets such as 25 percent cut by 2020 and 40 percent by 2030. It can be said that consensus was successfully reached based on knowledge from experts.

On the other hand, it is not quite sure whether consensus among private sectors (e.g., business, citizens) and policy makers can be reached, because basically only a limited number of selected members from private sectors are invited to committees and assembly meetings. Usually a wider range of private stakeholders is only informed by public events, newsletter, and public comment, etc.

The next step is to organize stakeholder dialogues to reach consensus and understand each other's role in developing sustainable (low carbon) societies at the local/regional/national level which are compatible with their preferred future. This lack of consensus seems to be the reason why concrete LCS actions rarely take place even though they have an excellent roadmap.

It is recommended that a one-day stakeholder dialogue called Focus Group Discussion (FGD) or a five- to seven-day workshop for main stakeholders called

Charrette should be organized to make a public plan in Malaysia. FGD and Charrette introduce techniques to encourage participants to perceive the problem as their own (ownership), and take responsibility and find a way to solve it (commitment). We believe that it is important to develop a methodology for how to utilize experts' information such as LCS scenarios and roadmaps for consensus building.

My next challenge is to develop processes based on scientific findings that a wider range of people can apply to achieve sustainable societies, and to help people join the decision-making process with their ownership and commitment.

References

"2050 Japan Low-Carbon Society" scenario team (2008), "Japan Scenarios and Actions towards Low-Carbon Societies (LCSs)," National Institute for Environmental Studies (NIES), Kyoto University, Ritsumeikan University, and Mizuho Information and Resarch Institute (MHRI), June. Online, available at: http://2050.nies.go.jp/report/file/lcs_japan/2050_LCS_Scenarios_Actions_English_080715.pdf (accessed August 29, 2011).

"2050 Japan Low-Carbon Society" scenario team (2009) "Japan Roadmaps towards Low-Carbon Societies (LCSs)," NIES, Kyoto University, Ritsumeikan University, and MHRI, August. Online, available at: http://2050.nies.go.jp/report/file/lcs_japan/20090814_japanroadmap_e.pdf (accessed August 29, 2011).

AIM (Asian Pacific Integrated Model) Project Team (2011) Online, available at: http://www-iam.nies.go.jp/aim/index.htm (accessed July 1, 2011).

Fujino, J. (2010) "Keep the Summit in Sight at COP16," *AJISS-Commentary* (Association of Japanese Institutes of Strategic Studies), 106(26), November. Online, available at: www.jiia.or.jp/en_commentary/201011/26–1.html (accessed August 29, 2011).

Kubo, H. (2011) "The Use of Expert Knowledge in Climate Change Policy: A Case Study of the Decision-making Process for Mid-term Reduction Targets," *Environmental Research Quarterly*, 16, 201–18 (in Japanese).

MOEJ (Ministry of Environment, Japan) (2011) "Mid-and Long-term Roadmap for Global Warming Measures." Online, available at: www.challenge25.go.jp/roadmap/index_en.html (accessed August 29, 2011).

NIES (National Institute for Environmental Studies) (2011) "Low-Carbon Society Scenarios towards 2050." Online, available at: http://2050.nies.go.jp/ (accessed July 1, 2011).

Nishioka, S. (2011) "Mid- and Long-Term Roadmaps for Global Warming Measures," International Symposium "Learning the Way to Building Low Carbon Society," Tokyo. Online, available at: www.challenge25.go.jp/roadmap/symposium_report_en.html#keynote (accessed August 29, 2011).

Research Team of Sustainable Society Kyoto (2009) "A Roadmap towards Low-Carbon Kyoto," August. Online, available at: http://2050.nies.go.jp/report/file/lcs_japanlocal/kyotolcs.pdf (accessed August 29, 2011).

Shiga Prefecture (Global Warming Prevention Office, Environmental Policy Division, Department of Lake Biwa and the Environment, Shiga Prefecture) (2011) "Global Warming Prevention Office." Online, available at: www.pref.shiga.jp/d/new-energy/ (accessed July 1, 2011) (in Japanese).

Shiga Prefecture Environmental Council (2010) "Towards Low Carbon Society in Shiga Prefecture," March 10. Online, available at: www.pref.shiga.jp/d/new-energy/teit-annso/tousin_d.pdf (accessed August 29, 2011) (in Japanese).

Shiga Prefecture Sustainable Society Research Team (2007) "Shiga's Scenario towards the Realization of a Sustainable Society," March. Online, available at: www.lberi.jp/root/jp/01topics/scenario_eng.pdf (accessed August 29, 2011).

Shiga Prefecture Sustainable Society Research Team: Roadmap Committee (2009) "A Roadmap for Sustainable Shiga Towards 2030," November. Online, available at: www.lberi.jp/root/jp/31kankou/3115panphlet/pdf/RoadmapE.pdf (accessed August 29, 2011).

Suda, M. and Fujino, J. (2010) "Analyzing the Implementation of Low-Carbon Society Study in Local Government: Case Study of Shiga Prefecture," *Proceedings of the Conference on Energy, Economy, and Environment*, 27, Tokyo, January 26.

Universiti Teknologi Malaysia (UTMI), Kyoto University, Okayama University and Ritsumeikan University (2009) *Low-Carbon City 2025: Sustainable Iskandar Malaysia*, Tsukuba, Japan: National Institute for Environmental Studies.

COMMENTARY TO PARTS I AND II

"Not such strange bedfellows: climate mitigation as a buttress for sustainable development"

Sandra S. Nichols, John A. Pendergrass, and J. Cory Connolly

Economic development is the main priority for developing countries. Standard economic development approaches rely on emission-intensive industrial technologies and practices. But on the international stage today there is a global imperative to reduce the emission of gases that are inexorably changing the climate. While the primary onus to reduce emissions is on industrialized countries because of their role in creating the problem, all nations are faced with the effects of climate change. In many cases, developing countries are the most vulnerable to threats from changed weather patterns, increased intensity of storms, and sea level rise. Developing countries, including those in Asia, are generally averse to accepting binding international commitments to reduce emissions. Nonetheless, efforts to mitigate climate-changing emissions can offer opportunities for funding and capacity building that may be appealing to developing countries. The availability of tools and approaches that allow climate mitigation efforts to support rather than undermine sustainable development can make these opportunities all the more likely to be pursued. Efforts to develop such tools and approaches have revealed that greenhouse gas (GHG) emissions and development are not as rigidly linked as once thought (Sathaye *et al.*, 2007).

Not only are developing countries threatened by climate change, but they are also progressively a bigger part of the problem. In pursuing economic development activities, they are increasingly contributing to global GHG emissions. And, as their climate-change-causing emissions increase, so do the threats from climate change. The UNFCCC process has focused on mandatory emissions reduction targets for industrialized countries (Annex 1). Developing countries (Non-Annex 1) have been encouraged to take voluntary actions. This approach has failed to meaningfully curb emissions. Industrialized nations are not taking strong enough steps to meet reduction targets and developing countries are not adopting voluntary actions that are significant enough to achieve meaningful results because many of the proposed approaches undermine economic development goals. Thus, curbing developing countries' emissions while maintaining or even enhancing development efforts has emerged as the greatest challenge in the international climate negotiations.

This book demonstrates that climate change mitigation and international development are not necessarily mutually exclusive goals. The case studies demonstrate the willingness of many countries to embrace voluntary mitigation actions – when climate mitigation and development efforts are mutually reinforcing. As efforts for broad and sweeping international action on climate change have stagnated, the case studies provide some hope by highlighting individualized approaches that address climate and development goals simultaneously.

One approach that serves as an appropriate and noteworthy frame for viewing these efforts is Development Policy Lending (DPL), examined in Chapter 8 by Taro Katsurai and Yuka Murakami. DPL enables developing countries to lead the way in identifying how they will reduce emissions – and how they will receive international support to do so. DPL inverts the standard approach of inserting development considerations into mitigation efforts by integrating mitigation efforts into development initiatives. Through policy decisions, Brazil, China, India, Mexico, South Africa, and Turkey have avoided 300 million tons of CO_2 per year over the past three decades (Sathaye *et al.*, 2007). These examples demonstrate that development and climate change mitigation can be reconciled, and that policy is an important tool in this process.

In the absence of significant unified action internationally, voluntary mitigation actions, like nationally appropriate mitigation actions (NAMAs) and the Clean Development Mechanism (CDM), have become the focus for reducing emissions on a regional, national, and local level. In 2008 there were only two CDM projects in Vietnam; in 2009 the total jumped to 56 registered projects, as described by Katsurai and Murakami in Chapter 8. Ling describes China and India's commitments to 40–45 percent and 20–25 percent reduction in carbon intensity by 2020, respectively. India has also adopted a domestic emissions trading scheme called Perform Achieve Trade (PAT), while Indonesia and Vietnam have undertaken policy based climate mitigation efforts and Bangladesh has adopted rural applications of solar technology.

Such voluntary efforts are not strictly for the global good; these countries have justifiable reasons for reducing emissions that align with national interests. Asia has a large carbon footprint and, simultaneously, is a region that will be disproportionately impacted by climate change. Foreseeing inundated coastal cities and declining crop yields, governments in this region are taking action. For example the majority of Bangladesh's population lives at 10 m or less above sea level (Angus *et al.*, 2009). The country has already seen increased flooding in recent years. Circumstances in Bangladesh are extreme, but other countries in the region are facing similar challenges.

Development and climate change are intertwined and climate change impacts are likely to aggravate and negatively impact development goals. Voluntary mitigation efforts can help avoid future harm from climate change, but they can also provide other benefits. When development objectives are advanced by mitigation efforts, these additional benefits are considered "co-benefits." Co-benefits result from the mitigation action, but are not the main goal. For example, improving public health is a co-benefit of reducing emissions in China. The emission intensive practices that have driven China's rapid industrialization have also caused air pollution that is affecting public health. Thus, emissions reductions would mitigate both GHG emissions and air pollution. Mitigation efforts that are properly aligned with development needs can produce co-benefits that improve local living conditions, enhance ecosystem services, meet livelihood needs, and provide economic opportunities.

Asian nations are making great strides in development and application of mitigation and adaptation strategies. Taking action on climate change can improve a country's international image, its access to international finance, and its political capital internationally. Some countries have made economic opportunity out of these efforts. China and India have become major players in the international market for low carbon technologies. China is rapidly dominating the solar panel manufacturing industry and both countries are developing other advanced clean energy technologies.

Ling points out that Asia currently emits 35 percent of greenhouse gases, so it is clear that Asia has a pivotal role to play in slowing climate change. Japan's role in the region also makes the Asian experience unique. As Fujikura highlights in Chapter 1, despite the disaster associated with the March 11, 2011 tsunami, Japan has demonstrated a strong commitment to accomplishing regional development and climate mitigation goals. This regional cooperation combined with a focus on policy-based strategies – like DPL – could help propel Asian developing countries toward a low carbon society. The efforts occurring in Asia are exemplary, but without enhancing these mitigation efforts and integrating them with development goals, limiting climate change or achieving sustainable development will be difficult. The examples in this book offer a valuable jumping off point for developing models that can be effective worldwide. Here, we suggest a set of considerations that could help build on these approaches for integrating development and climate mitigation.

Enhancing and deploying strategies for mitigation and sustainable development

Developing countries are working to reduce emissions, but sustainable development and economic growth remain their primary focus. Successful initiatives in Asia and elsewhere are proving that it is possible to reduce GHG emissions while continuing to prioritize development, even in the absence of firm emissions caps. These models can play an even more important role if they are developed appropriately for the recipient countries, are expanded to relate to sectors beyond energy production, and are developed and implemented at appropriate scales.

Policy-based approaches will build legitimacy

First, and most critically, a policy-based approach will vastly improve efforts to mitigate climate change in the developing world. Developing countries have rejected the "target-based approach" because binding commitments to emission reduction targets do not give them the flexibility they need to pursue economic development. DPL and similar policy-based strategies provide funding to support policy and institutional reforms that can address climate objectives and other objectives of the recipient country, such as sustainable development. Policy-based lending allows developing countries to integrate mitigation efforts into efforts such as reducing hunger and poverty, electrification, economic growth, establishing equal rights, and improving health. This "policy-based track" for reducing GHG emissions avoids lender-dominated policy development allowing the recipient country to design the programs. Any initiative that does not have legitimacy in the eyes of the recipient country is unlikely to result in long-term progress. DPL has the advantage of incorporating the principle of local ownership, which is the most successful way to establish and maintain the legitimacy of a project or policy.

The independence that DPL grants to recipient countries is demonstrated by the fact that funds provided are for general budget support that the central government can spend as it chooses. This policy-based approach is described in this book as it is implemented by JICA, but it has also been adopted, to some degree, by the World Bank, United States Agency for International Development (USAID), and Official Development Assistance through the Organization for Economic Cooperation and Development (OECD). Unlike international lending strategies of the past, DPL ensures developing countries play the dominant role in determining where and how assistance is used.

Katsurai and Murakami present the example of Indonesia's Climate Change Program Loan (CCPL). The CCPL used the DPL approach to support Indonesia in developing a comprehensive policy matrix that supports both climate mitigation and adaptation, focusing on the aspects determined most important by Indonesia. The authors also present Vietnam's experience with policy-based lending. In Vietnam, the government has built upon steady climate mitigation and adaptation efforts dating from its ratification of the UNFCCC in 1994, to greatly expand its registered CDM projects. The Special Program to Respond to Climate Change (SP-RCC) helped to build institutions, develop a series of policies and regulations, and served as the platform for communication and coordination that resulted in 56 new CDM projects. Communication and coordination were especially important in this effort because climate is such a completely cross-cutting issue. The SP-RCC was integrated into a pre-existing Vietnamese government program with related objectives. No new government institutions were required.

One element of the international debate about which countries should bear the burden of reducing emissions arises from the fact that mitigation has been promoted as the most important aspect of assistance to developing countries for some programs. DPL approaches the problem from a different perspective. It allows recipient countries to reverse the presumption – instead of climate change mitigation being primary, sustainable development becomes the focus, with climate considerations ancillary. Mainstreaming climate considerations into development practices is the most comprehensive way of addressing emissions reductions. Countries in Asia and the developing world will continue to prioritize development – ensuring that mitigation is a component of development programs will result in many more opportunities for emissions reduction efforts. The DPL models demonstrated in Asia offer valuable lessons for the rest of the developing world and those working in it. This valuable dynamic can be effectively used in the development of nationally appropriate mitigation actions (NAMAs). NAMAs vary based on national circumstance and will likely be central to any future international mitigation strategies.

While DPL incorporates important principles and offers hope for better integrating sustainable development and climate change mitigation, it does carry the legacy of structural adjustment lending, which was widely criticized for compelling countries to change fiscal policies in a way that was not always beneficial for their populations. Implementers of DPL projects should be vigilant to ensure that policies and projects are truly developed by, or at least jointly with, the recipient country and designed to accomplish the intended goals of sustainable development and climate mitigation.

Co-benefits facilitating development

Moving beyond existing strategies will only happen if development becomes a more prominent aspect of mitigation actions. Developing countries will only participate in policy-based efforts if they include appropriate co-benefits. Development related co-benefits are the reason why many countries are voluntarily taking action. In China, reducing air pollution is an important short-term benefit that accompanies efforts to improve the efficiency of coal-powered fire plants (Balbus *et al.*, 2009). As the Chinese reforest their land to sequester carbon, they also reduce erosion that otherwise would negatively influence ecosystems, local infrastructure, and, in turn, safety. India has a commercial fleet of compressed natural gas vehicles that are reducing pollution and, simultaneously, increasing energy security. In Argentina, urban forestation efforts are helping to remove carbon from the atmosphere and regulating micro-climates – an increasingly important adaptation co-benefit as global temperatures rise (LJ Ramos, 2010).

REDD + is a valuable example of an expansion of the international carbon markets that has inherent co-benefits. For example, in avoided deforestation scenarios, carbon is stored, ecosystems and ecosystem services are preserved, and communities, farmers, and others receive stable financial benefits. In Chapter 3, Yamada and Fujimori suggest that co-benefits be included in the scoring methodology for crediting in the CDM. There are already some existing approaches for voluntary market scoring methodologies that ensure carbon reductions and additional community-based co-benefits that can drive development. The Climate,

Community and Biodiversity Alliance and Plan Vivo have developed some important models (CCBA, 2011; Plan Vivo, 2008). Yamada and Fujimori's recommendation should be extended to any future international market mechanisms addressing climate mitigation. An emphasis on co-benefits in methodology scoring will better facilitate the accomplishment of contextually appropriate development objectives in addition to mitigation.

Developing a more inclusive Clean Development Mechanism

The CDM was designed as the Kyoto Protocol's method of fostering both sustainable development and mitigation. Currently, the CDM is one of the few international mechanisms that helps facilitate investment from developed countries to developing countries for climate mitigation. The CDM model allows a given project, such as renewable energy installations, energy efficiency measures, or the sequestration of carbon, to receive compensation for emissions reductions after they have occurred. The project-by-project nature of the CDM, however, makes for disjointed efforts at mitigation that are not designed to, and thus rarely, blend well with domestic or international development plans or strategies. These separate efforts are also unlikely to result in using resources as efficiently as possible. Moreover, many countries lack the technical knowledge, domestic resources, and financing to pursue CDM projects at all.

As a result of these challenges, the CDM has been less successful than anticipated. Only a small proportion of eligible countries have used the CDM. The current capacity of the CDM to mitigate emissions is vastly outpaced by the growth in emissions from developing countries. By the end of 2012, 400–600 $MtCO_2$ are expected to have been mitigated from the electricity sector thanks to the CDM, while electricity related emissions from Non-Annex I countries have grown by 60 $GtCO_2$ in the last decade (Aasrud *et al.*, 2010). On the whole, CDM is being used most by countries that already have strong economic growth. Currently, 75 percent of CDM projects are in China, India, Brazil, and Mexico (Dechezlepretre *et al.*, 2009). Reducing emissions in rapidly emerging countries is important, but to meet global emissions reduction goals, integrated efforts at mitigation and development must be available more universally.

There are several steps that can be taken to improve upon the experience to date with the CDM. Introducing broader market mechanisms to CDM efforts would enable the program to concurrently facilitate development and mitigation assistance. International carbon markets could be expanded by streamlining methodology development, improving access to information and guidance, and developing broader methodologies that can be applied to a wider range of projects dealing with a specific technology or sector. For example, the CDM would become more broadly relevant if certified emission reduction credits were offered for Non-Annex I country actions that resulted in emissions reductions above and beyond NAMA commitments (Aasrud *et al.*, 2010). This would also complement and support domestic policy-based actions and decisions on climate by providing positive financial incentives for over-compliance and no penalties for not meeting mitigation goals.

In Chapter 3, Yamada and Fujimori highlight the need for developing a system for funding projects that have already been accepted by the CDM. Additional financial resources are also needed for projects that are in the development stage. Without initial assistance or financing, many projects that would open up economic opportunities to different segments of society would never be developed. The CDM can only reach its potential if there is support to experiment with a range of approaches. Many countries will not be able participate without support from the start. The expansion of the CDM or the advent of other international market mechanisms will be critical for reducing emissions from more actors and sectors. To be most effective in addressing mitigation and development, host countries should play a larger role in shaping new or expanded market mechanisms.

There are many examples of the opportunities presented by new technologies. Consistent with the idea of nationally appropriate mitigation actions and the expansion of CDM projects, in Chapter 10 Kaneko, Komatsu, and Ghosh point out the need to employ context appropriate technologies, not just proven technologies. Increased investment in experimental and innovative technological solutions in the developing world will lead to advancement of technology in a manner that is contextually appropriate. A policy-based approach that enables developing countries to control how international financing and support is applied and allocated will help ensure that technologies, like solar power generation, are properly designed to meet mitigation goals and electrification development needs. Bangladesh has deployed solar technologies that will "leap-frog" traditional energy infrastructure. South American counterparts like Bolivia, Peru, and Ecuador have also made significant investments in solar hot water and solar photovoltaic technologies with European support (Eurosolar, 2011). These investments will help move past the restrictions of traditional grid infrastructure for rural populations. More investments in experimental technologies, whether through NAMAs or DPL, will lead to increased efficiency in mitigation efforts, technology, and development.

Nationally appropriate mitigation actions

As the developing world becomes more involved in mitigation efforts and in international carbon markets – whether through the CDM or other market mechanisms – actions will have to be contextually appropriate. NAMAs are increasingly the vehicle for developing countries to reduce emissions. First introduced as a part of the Bali Action Plan at COP13 in Bali as voluntary actions, NAMAs are developed by a non-Annex 1 country for its specific needs and are non-binding. Domestic policies like feed-in tariffs, technology performance standards, the application of market-based instruments, and a variety of other policies and activities can be used at the domestic level as a part of a NAMA. As Yamada and Fujimori point out in Chapter 7, NAMA strategies could be enhanced by including opportunities for indirect GHG reductions that might occur through institutional and educational activities. Sharing information between developing countries would also be useful for advancing these disaggregated approaches to mitigation. Nationally appropriate strategies are consistent with the policy-based approach that incorporates development and mitigation. These individualized strategies allow national governments to decide which activities are most cost effective and will most appropriately align with their development goals. Providing international incentives, whether through NAMA crediting or some other mechanism for policy-based mitigation action will allow for more innovation on the part of developing countries and will encourage more ambitious mitigation action in the future.

The conditions and circumstances of every country differ, so strategies adopted by various countries will vary. In many developing countries, Toyota points out in Chapter 9, coal-fired power plants are a necessity. Inefficient coal-fired power generation is one of the greatest contributors to GHG emissions, but for developing countries where electrification is not yet widespread, coal power is an important vehicle for achieving development goals. Many coal-fired power plants use old technologies that are very inefficient. While many would prefer to see the complete elimination of coal-fired power plants, for countries seeking electrification, coal may be the only reasonable alternative in the short to medium term. But this decision is not a binary one. Higher efficiency power plants offer the opportunity for a compromise. Providing credits or financing for higher efficiency coal power plants – although unpopular in many circles – would help lower emissions. Another benefit of improving coal-fired power plant efficiency is to reduce local pollution and the presence of black carbon (or soot) in the atmosphere. Black carbon accounts for 10–20 percent of global warming and is prevalent in developing countries where cleaner and more efficient plants could be deployed (Baron *et al.*, 2009).

Providing credits for efforts like constructing more efficient coal-fired power plants to reduce emissions could also be incorporated into a sector or NAMA-based crediting system or an expanded CDM. Such a system could allow for more autonomy in determining how emissions will be reduced – for example, giving countries the option of choosing more efficient coal-fired power generation – while maintaining levels of economic growth and development.

Nationally appropriate strategies foster mitigation efforts in the developing world, but there are potential pitfalls. Any mitigation project needs to start with a detailed assessment. This is particularly important for policy-based projects because they will have broad implications for the country. As pointed out by Ling in Chapter 6, if a detailed assessment is not undertaken then mitigation actions may not be as effective as intended or they could be wasteful. For example, incentives for using new technologies must be properly aligned with the demand. India's policy framework reduced up-front costs to wind energy developers with the result that India is now the third largest wind energy installer. But, because the incentives were for development and not generation, many of the plants are operating at less than their full capacity. Even more worrisome is the simple fact that NAMAs are non-binding. Most trading schemes rely on the existence of binding emissions caps; developed countries may balk at such caps, as has the United States, unless the emerging economies adopt them as well. These binding emissions trading schemes in developed countries are critical to limiting global warming, and any future international crediting mechanisms would likely rely heavily on the demand generated by the continuation of a trading scheme in the developed world – without binding emissions schemes in the developed world, NAMA crediting, sectoral crediting, and CDM efforts become components of an unstable voluntary market.

Sectoral approach to rendering actions appropriate

This book demonstrates some very important initiatives and lessons, but the case studies focus on the energy and waste sectors. Although these sectors are significant contributors to GHG emissions, there are important examples from other sectors including transportation, housing, and urban planning, that also demonstrate opportunities for integrating development and emissions reductions. While the case studies in the book present approaches to improve efficiency in energy and waste sectors, these other sectors are doubly important because they provide an opportunity to reduce the need for energy. Focusing on the energy sector might not always be the most appropriate for each country. In fact, a sectoral approach is likely to be most effective because it provides the opportunity to tailor programs to conditions in each country and will likely address more sectors than just energy and waste.

This book offers an important look at mitigation options in rapidly developing Asia. In Chapter 6, Ling discusses sectoral approaches for China and India in energy, industry, transportation, and buildings. Other case studies focus on more efficient coal-fired power plants, distributed solar systems, and waste management, and other renewable energy solutions. Emissions limiting or reducing policies will apply to different sectors depending on the domestic context. In many countries, particularly in the developing world, a countrywide policy approach that incorporates many sectors is not politically, economically, or practically feasible. The Indian emissions program, PAT, includes nine sectors (Aasrud *et al.*, 2010). In Mexico, the government has targeted five specific sectors (Johnson *et al.*, 2009). These approaches may be broader than is feasible for many developing countries to undertake. Thus, it will be important to develop approaches that can be applied to particular sectors. The sectors that make the most sense will vary on a country by country basis.

While the energy sector gets the most attention and the waste management, transportation, and the building sectors are often identified as the most manageable, urban planning, land use and agriculture, and other sectors are also important (Sathaye *et al.*, 2007). Sectors such as building, urban planning, transportation, and agriculture are significant energy users and contributors to GHG emissions. Potentially more importantly, these other sectors have major implications for development as well.

Much attention has been paid to sector specific market mechanisms in efforts to limit emissions post-2012. Sectoral crediting, where an emissions baseline would be established for a sector and credits rewarded for operating at emissions levels under the baseline, is a possible means for reducing emissions in developing countries. Sector policies can also be a component of NAMAs. The most efficient and feasible sectors to address will depend on circumstances in each country – another reason why a policy-based approach may be the preferred approach for developing countries with different sectoral priorities.

While the priority of sectors will vary from country to country, it is also important to recognize the difficulties in addressing only one or two sectors when many of the sectors are interdependent. Urban planning is a particularly good example. Effective urban design will help reduce energy usage, improve efficiency of transportation, and effectively manage waste. City-to-city collaboration through existing and new avenues can support sustainable low-emission urban planning efforts, but such efforts will also have implications for other sectors. In developing countries where capacity and resources for mitigation may be minimal, undertaking efficiency improvements in high priority sectors will most effectively contribute to development and mitigation goals.

From the international to the local

Finally, while the technical and policy considerations described above are necessary, an additional consideration for policy makers is the scale at which emissions reduction efforts are implemented. International development and mitigation efforts are often viewed through the lens of international institutions. International agreements, organizations, and forums are seen as the primary methods for international action to take place; however, local, regional, and bi-lateral initiatives are increasingly playing a role. Driven in the last few decades by the imperative of climate change mitigation and adaptation, highly effective networks and organizations have developed that facilitate local-level action on climate change; as Nakamura and Mori underscore in Chapter 5. Cooperation and coordination at the regional, national, and municipal level can enhance the effectiveness of mitigation strategies, while also pursuing international development goals. City-level action can also provide a foundation for innovative and broader policy and market mechanisms. Rather than considering only NAMAs, it is also important to examine such actions that are also locally appropriate.

Intercity and regional collaboration has become more and more important as national and international methods for decision making and policy development have been, at times, slow or stagnant. In Asia, successful intercity collaboration has been seen through organizations such as ICLEI, the Kitakyushu Initiative, CITYNET, and CAI-ASIA (see Chapter 5). These organizations help facilitate collaboration between local governments. As Nakamura and Mori explain, when given a platform for collaboration and the autonomy to do so, cities can provide lateral technical support, share best practices, and take action that might not be achievable at other levels of government. In China, such city-level action has taken the shape of city-level emissions trading schemes in Shanghai, Bejing, and Tianjin (Aasrud *et al.*, 2010).

Grassroots efforts blossoming into intercity and local government action can be found in many places beyond Asia. Organizations and initiatives such as C-40, Urban Sustainability Directors Network (USDN), and the Joint US/China Collaborative on Climate Change, all are helping facilitate action at the local level. C-40 is a partnership between the Clinton Climate Initiative and the Large Cities Climate Leadership Group in order to help reduce emissions in large cities throughout the world. This partnership recognized that cities emit as much as 70 percent of GHGs and thus must become significant actors in efforts to reduce emissions (C-40, 2011). Municipal members of the initiative are developing policies and projects for reducing GHG emissions appropriate to their specific needs, but are also sharing their strategies among the other members as well as publicly. For example, New York has developed a comprehensive climate change plan and is working toward transitioning its entire fleet of taxis to hybrid vehicles. Another example is the USDN, predominantly functioning in the United States, which facilitates direct collaboration between city officials (USDN, 2011). C-40 and USDN demonstrate that direct interaction between city officials can be effective in helping develop local policies that promote development and mitigation worldwide. The International Council for Local Environmental Initiatives (ICLEI), on the other hand, takes a more involved role in facilitating city and local government action. ICLEI offers services, conferences, and tools to help facilitate sustainability – including adaptation and mitigation practices. City and local government collaboration is also occurring between Asian countries and the United States. The Joint US/China Collaborative on Climate Change is facilitating an exchange of information and practices from one part of the world to the other. Providing opportunities and incentives to cities to tackle climate change and helping to facilitate intra- and intercity exchanges will help produce more solutions that are locally appropriate for mitigation and development. Such lateral initiatives and partnerships also have the potential to facilitate bidirectional exchange of best practices and expertise – meaning that the developing world could, and likely will, increasingly provide innovative ideas and strategies for addressing climate change in the developed world.

Cities throughout the world have made voluntary commitments, taken voluntary mitigation actions, and, in some instances, have even engaged in the international carbon market through CDM projects. At the climate change negotiations in Cancun, considerable discussion focused on city-level action, particularly CDM efforts in Amman, Jordan. Amman had proposed a citywide CDM project that would incorporate a variety of sectors under one project. The city of Amman, C-40, and ICLEI, collaborated to develop language that was ultimately inserted into the CDM draft decision coming out of Cancun. The draft specifically mentioned future "city-wide programs" (UNFCCC, 2010). With the citywide methodology awaiting approval from the CDM Executive Board, it is unclear how cities will ultimately fit into the CDM. Whether approved or not, the inclusion in the draft decision indicates that it is important for future market instruments to include cities and citywide mitigation efforts.

In the future, local efforts will likely become even more important in developing mitigation and development solutions. In an urbanizing world, cities already account for the vast majority of emissions. Many emissions reductions efforts can only be effectively carried out at the local level. Developing cities in a more sustainable manner will reduce emissions and provide development benefits including more efficient transportation and improved air quality.

Conclusion: global climate solutions require voluntary developing country action

The prospects for a renewed international climate agreement are not particularly positive in the near term. After the current commitment period of the Kyoto Protocol expires in 2012, the trend in climate change response may shift to the national and local level. In a more disjointed post-2012 international climate policy landscape, mainstreaming mitigation into development practices will be critical if the world hopes to limit global temperature rise to 2 degrees Celsius. Actions in Asia demonstrate that mitigation and development can be pursued in tandem. Mitigation efforts will have to adequately guarantee co-benefits that contribute to economic development, although "co-benefit" may be the wrong term for such benefits, or perhaps its application should be reversed to refer to mitigation outcomes achieved through development. In much of the developing world, economic development is primary and mitigation is secondary. The centrality of mitigation in the "mitigation/development" equation, as reflected in international debates, is not appropriate in the developing world. For developing countries, development and adaptation take priority, which is why integrating mitigation into development and adaptation strategies will be increasingly important in the future.

The world has the tools needed to support developing countries in mitigating climate change, but without expanding and replicating the most effective policies and projects adopted by developing countries, emissions reduction goals will not be met. The examples in this book along with the resulting lessons learned can serve as important experiences for other regions. In particular, targeting a wider range of sectors and engaging international action at multiple levels of governance are important lessons. Additionally, the examples of emissions mitigation approaches presented can be informative in efforts to expand international carbon markets and better integrate development and mitigation. A policy-based approach for the developing world seems the most apt to support the actions already taking place in Asia and to be applied more broadly. It has significant potential as a vehicle to, as the introduction to this book terms it, a "low carbon society."

Efforts to combat climate change will not be effective if the nations of the world do not work together to build on experiences and support each other in efforts to reach these goals. This book should serve as an example of how to share lessons learned. Additionally, as noted in the introduction, Japan's effort to rebuild can help set an example and provide resources as other countries seek mitigation strategies. Japan's presence in the region and its current shift away from nuclear power after the 3/11 tsunami create an opportunity for close collaboration between developed and developing countries in Asia on mitigation, development, and rebuilding. This spirit of collaboration provides a strong example of an alternative approach, which can set a tone for the rest of the world. Experiences such as those described here must continue to be shared and lessons transferred as IPCC negotiations continue and at international venues such as upcoming Rio+20.

JICA has now produced two informative books on responses to climate change and international development: *Climate Change Adaptation and International Development: Making Development Cooperation More Effective* and this book *Climate Change Mitigation and International Development.* With limited climate financing available and with development concerns continuing to be primary, it will be important to examine the synergies between adaptation, mitigation, and development. To continue to improve and promote mechanisms and strategies for mitigation actions, research into potential synergies between mitigation and adaptation in order to reduce costs and protect developing countries from climate impacts should be pursued.

References

Aasrud, Andre, Richard Baron, and Katia Karousakis. (2010). "Market Readiness: Building Blocks for Market Approaches." IEA and OECD. Online, available at: www.oecd.org/dataoecd/30/14/46563135.pdf (accessed September 13, 2011).

Angus, S.D., B. Parris, and B. Hassani. (2009). "Climate Change Impacts and Adaptation in Bangladesh." Department of Economics, Monash University and World Vision Australia, July. Online, available at: www.mssanz.org.au/modsim09/G4/angus.pdf (accessed September 13, 2011).

Balbus, John, Jacob Bell, and Sarah Penney. (2009). "Estimating the Health Impacts of Coal-Fired Power Plants Receiving International Financing." Environmental Defense Fund. Online, available at: www.edf.org/sites/default/files/9553_coal-plants-health-impacts.pdf (accessed September 15, 2011).

Baron, Richard, David Montgomery, and Sugandha Tuladhar. (2009). "An Analysis of Black Carbon Mitigation as a Response to Climate Change." Copenhagen Consensus Center, Washington, DC. Online, available at: http://fixtheclimate.com/uploads/tx_templavoila/AP_Black_Carbon_Baron_Montgomery_Tuladhar_v.4.0.pdf (accessed September 19, 2011).

C-40, Climate Leadership Group. (2011). "CDP 2011: Global Report on C40 Cities." Carbon Disclosure Project. Online, available at: http://c40citieslive.squarespace.com/storage/CDP%20Cities%202011%20Global%20Report.pdf (accessed September 17, 2011).

CCBA, Climate, Community, and Biodiversity Alliance. (2011). "Scoring Summary." Online, available at: www.climate-standards.org/standards/scorecard.html (accessed September 19, 2011).

Dechezlepretre, Antoinne, Matthieu Glachanta, and Yann Meniere. (2009). "Technology Transfer by CDM Projects: A Comparison of Brazil, China, India, and Mexico." *Energy Policy*, 37, 2, pp. 703–711.

Eurosolar. (2011). "Eurosolar: Renewable Energy for Development." Online, available at: www.programaeuro-solar.eu/eng/home.php (accessed September 19, 2011).

Johnson, Todd M., Claudio Alatorre, Zayra Romo, and Feng Liu. (2009). "Mexico: Estudio Sobre La Diminucion De Emisiones De Carbono." Banco Mundial, 8 October. Online, available at: http://siteresources.worldbank.org/INTLAC/Resources/MEDEC_Executive_Summary_Spa.pdf (accessed November 21, 2011).

LJ Ramos Brokers inmobilarios. (2010). "Indice Verde: Informe Especial del Mercado Inmobiliario." September. Online, available at: www.ljramos.com.ar/upload/IMI_septiembre2010.pdf (accessed September 19, 2011).

Plan Vivo, Carbon Management and Rural Livelihoods. (2008). "Plan Vivo Standards: 2008." Online, available at: http://planvivo.org.34spreview.com/wp-content/uploads/Plan-Vivo-Standards-2008.pdf (accessed September 19, 2011).

Sathaye, J., A. Najam, C. Cocklin, T. Heller, F. Lecocq, J. Llanes-Regueiro, J. Pan, G. Petschel-Held, S. Rayner, J. Robinson, R. Schaeffer, Y. Sokona, R. Swart, and H. Winkler. (2007). "Sustainable Development and Mitigation." In *Climate Change 2007: Mitigation. Contribution of Working Group III to the Fourth Assessment Report of the Intergovernmental Panel on Climate Change* [B. Metz, O.R. Davidson, P.R. Bosch, R. Dave, L.A. Meyer (Eds.)], Cambridge University Press, Cambridge, UK and New York, USA.

UNFCCC. (2010). "Decision 3, Report of the Conference of the Parties Serving as the Meeting of the Parties to the Kyoto Protocol." Cancun, Mexico, November 29 to December 10, p. 3. Online, available at: http://unfccc.int/resource/docs/2010/cmp6/eng/12a02.pdf#page=2 (accessed September 19, 2011).

USDN, Urban Sustainability Directors Network. (2011). Online, available at: http://sustainablecitiesinstitute.org/usdn (accessed September 19, 2011).

INDEX

Note: Page numbers in **bold** are for figures, those in *italics* are for tables

Abe, Shinzo 12, 155, 228
Ad hoc Working Group on Further Commitments for Annex 1 Parties under the Kyoto Protocol (AWG-KP) 4, 5, 23, 24, 25
Ad hoc Working Group on Long-term Cooperative Action under the Convention (AWG-LCA) 23, 24, 25, 26
adaptation 26, *170*, 244
Afghanistan 122, *123*
Africa 45; deforestation 51; *see also* Congo; Ghana; Nigeria; South Africa; Sudan; Tanzania; Zimbabwe
Agence Française de Développement (AFD) 10, 156, 157, 161, 165, 169
agricultural sector 249; India 113; Indonesia *153*, *158*, *162*; Vietnam *167*
aluminium industry 105
Amman, Jordan 251
Annex I countries 4, 21, 22–3, 24, 37
Annex II countries 21
appropriate technology 224, 247
Argentina: Clean Development Mechanism (CDM) projects 43; deforestation *52*; forestation 245
Asia-Pacific Eco-Business Forum (2006) 79
Asia Pacific Economic Cooperation (APEC) 5, 33–4
Asian City Network for Environment Improvement 93
Asian Development Bank (ADB) 11, 12, 33, 65, 155, 210, 218
Asian Pacific Integrated Model (AIM) 227, 237
Aso, Taro 229
Association of Southeast Asian Nations (ASEAN) 122

Bachat Lamp Yojana program 112
back-cast models 228, 229
Bago 72
Baguio *69*, 70, 71
Bali Action Plan (2007) 4, 5, 8, 23, 24, 54, 120, 154, 247
Banda Aceh 68, *71*
Bangkok 4, 68, *69*, 70, 72, 81
Bangladesh 32, 202–26, 242; electrification (anti-corruption system 205–7; grid 203–4, 204–9, 219–20, 221; and household incomes 214, 215; income generation effects 219–20; international support 218–19; livelihood improvement effects 220; and mobile phone ownership 212–13, 214, 215; *Palli Bidyut Samity* (PBS) 11, 205, 206, 209; Performance Target Agreement (PTA) 205; solar home systems (SHS) 11–12, 203–4, 209–24, 242, 247); Infrastructure Development Company Limited (IDCOL) 11, 210, 211, 218, 219; kerosene consumption 214–15, 216; rechargeable battery ownership 214, 216, *217*; Rural Electrification Board (REB) 11, 204–5, 207, 209, 219

Bangladesh Power Development Board (BPDB) 204, 205, 207
BASIC group 19
batteries, rechargeable 214, 216, *217*
Beijing 250
Berlin Mandate 21–2
Bhutan, nationally appropriate mitigation actions (NAMAs) 122, *123*
bilateral cooperation 34, 67–8, 75, 181
Bilateral Offset Credit mechanisms 142
binary probit model 214
biodiversity 34, 58
biogas projects 43, *44*
biomass: burning 108; forest 55, 56
biomass power 43, *44*, 100, 109, 110
black carbon 248
Blair, Tony 23
Bohol Province *69*, 71
Bolivia 25, 247
bottom/base of the pyramid (BOP) market, for solar home systems 222, 223–4
BRAC Foundation 210
Brazil 242; BASIC group membership 19; Clean Development Mechanism (CDM) projects 5, 43, 246; deforestation 51, *52*; nationally appropriate mitigation actions (NAMAs) 122, *123*
Brunei 4
buildings sector 249; China 106; India 111–12; low carbon development policies *83*, *84*
business-as-usual (BaU) emissions 32, 120, 122, 203
business sector: development 78, 86, 91; low carbon development in 81, 87–8, 90, *92*, 93

C40 Cities 79, 117, 251
Cambodia 237; forest carbon stocks 56; nationally appropriate mitigation actions (NAMAs) 122
Canadian International Development Association (CIDA) 169
Cancun Adaptation Framework 26
Cancun Agreement (2010) 3, 5, 8, 25–7, 120–1, 149
capacity building 23, 47, *65*, *66*, 67, 81, 90, 116, 237
carbon capture and storage (CCS) 11, 35, 104, 114
carbon credits 81, 88, *92*, 180
carbon dioxide emissions *see* emissions
carbon intensity 10, 99, 100, 175, 185, *187*; India 108
carbon markets 22, 23, 246
carbon offsetting 7, 81, 88–9, *92*
Cebu *69*, 71
cement industry, India 111
Certified Emission Reductions (CERs) 5, 30, 37, 40, 41, 42, 43, *44*, 45
chemical industry 105
Chiang Mai 71
Chiba *82*
Chile, Clean Development Mechanism (CDM) projects *43*
China 5, 8, 11, 27, 101–6, 113, 114, 237, 242; BASIC group membership 19; buildings sector 105–6; Clean Development Mechanism (CDM) projects 43, 246; coal-fired power 102–3, 104, 114, *183*, 190, **191**, **192**; economic growth 19; emissions 8, 19–20, 99, **100**, 101; emissions reduction *31*, 102, 121, 122, *124*, 196, *197–8*, 199; energy consumption 8, 105, 106, 114; energy demand 8, 99, **100**; energy efficiency 29, 192, 195, 196; energy intensity 101, 102, 105; energy production 8, 102–4; energy security 102, 103, 106; energy supply structure 189, 190; Five Year Plans (FYP) 101, 102; forestation 51, 101, 245; industry sector 105, 243; National Climate Change Action Program (NCCAP) 101; National Development and Reform Commission (NDRC) 101; National Renewable Energy Law 103; nationally appropriate mitigation actions (NAMAs) 122, *124*; natural gas **191**, 192; oil-fired power **191**, **192**; renewable energy 113, 114 (hydropower 103, **191**; solar power 8, 103; wind power 8, 103, **104**); sectoral policy 101–6; solar panel manufacturing 243; Ten Key Energy Conservation Projects 105, 106; Top-1000 Energy-Consuming Enterprises program 105; transportation 106, 115; and US Collaboration on Climate Change 251
Chongqing 81
cities 250, 251, 252; eco-model 85, 234; *see also* intercity collaboration
Cities for Climate Protection (CCP) programme 71
Cities Development Initiative for Asia (CDIA) 62
CITYNET 7, 62, 63, 64, *65*, 67, 68, *69*, 71, 73, 93, 117, 250
Clean Air Initiative for Asian Cities (CAI-Asia) 7, 62, 63, 64, *65*, *66*, 67, 250

Clean Development Mechanism (CDM) 5, 11, 13, 22, 30, 57, 90, 126, 164, 180, 193–4, 245; and afforestation and reforestation (AR-CDM) 53–4; reform of 46–8
Clean Development Mechanism (CDM) projects 5–6, 15, 37–48, *143*, *144*, *159*, 242, 244, 246, 251; additionality 40–1, 45; approval of 39–40; baseline scenario 40; co-benefits 47, 48; critical issues of 45–6; current status of 42–4; developing and designing 38–9; financing 48, 247; monitoring of 41–2; procedures and key concepts 38–42; regional distribution of 45, 46–7; small-scale 42; unilateral 39–40; validation and registration of 40–1; verification and certification of 42
Clean Technology Fund (CTF) 115
climate change program loan (CCPL), Indonesia 151–63, 171, 172, 244
Climate Change Vulnerability Index 163
Climate, Community and Biodiversity Alliance 246
Climate Technology Center and Framework 27
Clinton Climate Initiative (CCI) 79, 251
co-benefits: Clean Development Mechanism (CDM) projects 47, 48; facilitating development 243, 245–6, 252
coal-fired power 10–11, 114, 185, 188–9, 203, 248; carbon intensity 10; China 102–3, 104, 114, **191**, **192**; India 106, 108, 110–11, **191**, **192**; Indonesia 152; Japan **192**; Malaysia *183*, **191**, **192**; ODA emissions reduction projects 193–201; thermal efficiency *190*, **192**, 194; Vietnam *183*, **191**, **192**
coastal erosion 27
collaboration *see* intercity collaboration
Colombo *69*, 70, *71*
compressed natural gas (CNG) vehicles 112, 245
Congo, deforestation *52*
construction industry, emissions **184**
Convention on Biological Diversity 58
"Cool Earth 50" 12, 155, 174n20, 228
Cool Earth Partnership 155, 181
cooperation: bilateral 34, 67–8, 75, 181; international 116; regional 32–4, 116; *see also* intercity collaboration
Copenhagen Accord (2009) 5, 8, 24, 25, 26, 31, 54, 180
corporate social responsibility (CSR) 88
Costa Rica 54
Creech, H. 76
crop yields 27, 28, 152

Dalian 71, *72*, 75, 78, 81
deforestation 30, 49, 50, 51, *52*, 53, 55, 58
demand-side management policies (DSM) 112
Dentsu Research 3
development policy lending (DPL) 9–10, 13, 149–79, 242, 244–5; Indonesia 151–63, 172; overview 150–1; Vietnam 163–71, 172
Dhaka 68, *69*, 70, *71*, 72
Dhaka Electric Supply Authority (DESA) 207
Dhaka Electric Supply Company Limited (DESCO) 207
DNA Forum 46

early warning systems, Indonesia *159*
Eco Model Cities 85, 234–5
eco points 88, 89, 95n32
Eco Town project (UNEP) 78
economic development 227, 241, 242, 244
economic growth 8, 19, 31, 151, 243; Indonesia 152; Vietnam 164
Ecuador 247
Egypt: Clean Development Mechanism (CDM) projects *43*; yen loan projects in power generation sector *200*
El Niño 152
electricity generation 10–11, 182, 185, 188–92; Bangladesh *see under* Bangladesh; per capita *186*
electrification rates 185, **188**
emissions 3–4, 10–11, 28, 29, 50, 99, 182, **202**, 203; business-as-usual (BaU) 32, 120, 122, 203; coal power generation 10–11, 108, 189; natural gas power generation 189; oil power generation 189; per capita 3–4, 19, 28, 100, 185, *187*; per unit of GDP 8, 31, 32, 33, 102, 121; by sector **184**; vehicle 112; *see also under individual countries*
emissions reduction 5–6, 11, 14, 24, 27, 88, 100, 122, 180, 227; indirect 13, 247–8; ODA coal-fired power plant projects 193–201; policy-based approach 172–3, 244–5; target-based approach 172, 244
emissions reduction targets 21, 29, 32, 121–2, *123–5*; Annex I countries 22–3, 26, 37, 241; Brazil 122, *123*; China *31*, 102, 121, 122, *124*; India *31*, 107, 122, *124*; Indonesia *31*, 122, *124*, 152, 161; Japan 12–13, *31*, 228, 229–31; Korea

emissions reduction targets – *contd.* *31*, 122, *124*; Mexico 122, *125*; Singapore *31*, 32, 122, *125*; Thailand *139*, *140*, *141*, 145
emissions trading 22, 180, 249, 250; EU 23; India 242
energy consumption: China 8, 105, 106, 114; India 8, 111, 114; Vietnam 164
energy efficiency 29, 34, 85, 100, 112, 190–2, 195–6
energy intensity 34, 99, 185, *186*; China 101, 102, 105
energy production: China 102–4; India 8, 108–11; Indonesia 164; Vietnam 164
energy security 5, 14, 29, 35; China 102, 103, 106; India 107, 108, 112
environmental education 7, 73, 88, 89
Environmental Law Institute 13
Espinosa, Patricia 25, 26
Eukuda, Yasuo 155
European Union (EU), regional emissions trading scheme 23
extreme weather patterns 32

feed-in tariffs 247
financial cooperation 64, 65, *66*, *67*
financing 25, 26–7, 30, 73, 115–16; Clean Development Mechanism (CDM) projects 48, 247; intercity collaboration 78–9; NAMAs 126, 142
Food and Agriculture Organization (FAO) 6, 50–1, 58
forestation: Argentina 245; China 51, 101, 245; India 51
forests: carbon stocks 54–7; degradation 49, 53, 54, 55, 58; emission sequestration 22; management and conservation 6, 25, 30, 34; Indonesia *158*, *162*; *see also* deforestation; REDD and REDD-Plus
fuel economy standards (FES) 106
Fukuda, Yasuo 162, 181, 228, 234
Fukuoka City *82*
Fukushima Daiichi nuclear power plant 1, 2

G20 process 23
G77 plus China group 19
G8 process 23
gas *see* biogas; natural gas
GDP, emissions per unit of 8, 31, 32, 33, 102, 121
Georgia 122
German Development Cooperation (KfW) 12, 210, 218, 219
German Technical Cooperation (GTZ) 12, 210, 218, 219
Ghana 122
glacier melt 32
Gleneagles Summit (2005) 23
Global Environment Facility (GEF) 115, 210
Global Partnership on Output-Based Aid (GPOPA) 12, 218
global temperature increase 3, 14, 24, 26, 99, 149
Gore, Al 23
Grameen Shakti 210–11, 219
Green Climate Fund 25, 27, 177n63
green procurement 79
greenhouse gas emissions (GHG) *see* emissions

Hanoi 68, *71*, 78
Hatoyama, Yukio 181, 229
HFC-23 decomposition 5, 43, *44*, 45
High Level Seminar on Environmentally Sustainable Cities (2010) 61
Hiroshima 85
Ho Chi Minh City 85
Hokkaido Prefecture *82*
households: low carbon development policies *83*, *84*; *see also* solar home systems
housing 249
Hue, Vietnam 68, *71*
hydrofluorocarbons (HFCs) 5, 43, *44*, 45, 180
hydropower 43, *44*, 100, 188; China 103, **191**; India 109, 110, **191**; Malaysia **191**; Vietnam 190, **191**, 199; yen loan projects *200*

ICLEI *see* International Council for Local Environmental Initiatives
Incheon 68, *71*
income(s) *186*; and electrification, Bangladesh 214, 215, 219–20
India 5, 8, 11, 106–13, 114, 242; agricultural sector 113; BASIC group membership 19; buildings sector 111–12; Bureau of Energy Efficiency (BEE) 111, 112; carbon intensity per capita 108; Clean Development Mechanism (CDM) projects 43, 246; coal-fired power 106, 108, 110–11, *183*, 190, **191**, **192**; compressed natural gas (CNG) vehicles 112, 245; Electricity Act (2003) 108; emissions 8, 99, **100**, 106, 107, 108; emissions reduction *31*, 107, 122, *124*,

152, 196, *197–8*, 199; Energy Conservation Act 111; Energy Conservation Building Code (2007) 111; energy consumption 8, 111, 114; energy demand 99, **100**, 106; energy efficiency 112, 192, 195, 196; energy production 8, 108–11; energy security 107, 108, 112; energy supply structure 189, 190; forestation 51; industry 111; National Action Plan on Climate Change (NAPCC), 2008 8, 107, 112; National Auto Fuel Policy 112; National Mission for Enhanced Energy Efficiency 111; National Mission on Sustainable Habitat 112; National Tariff Policy (2006) 108, 111; nationally appropriate mitigation actions (NAMAs) 122, *124*; natural gas **191**, **192**; oil-fired power **191**, **192**; Perform Achieve Trade (PAT) scheme 242, 249; poverty reduction 106, 107; Prime Minister's Council on Climate Change (PMCCC) 107; public-private partnerships 111; renewable energy 113, 114 (biomass power 109, 110; hydropower 109, 110, **191**; solar power 29–30, 109; waste-to-energy 109, 113; wind power 8, 109, 248); Renewable Energy Development Agency 109; renewable portfolio standards 109; sectoral policy 106–13; transportation 112–13, 115; waste management sector 113

Indo-German Energy Efficiency and Environment Project 111

Indonesia 9 10, 58, 237, 242; agricultural sector *153*, *158*, *162*; Clean Development Mechanism (CDM) projects *43*, *159*; climate change program loan (CCPL) 151–63, 171, 172, 244; coal consumption 152; crop production 152; deforestation 30, 51, *52*; development policy lending (DPL) to 151–63, 172; economic growth 152; emissions 152–4; emissions reduction targets *31*, 122, *124*, 161; energy demand 152; forest management *158*, 161, *162*; National Action Plan on Climate Change (NAP-CC) 9, 154, 156, 157; National Action Plan to reduce GHG emissions 161; National Council for Climate Change (NCCC); 161; National Development Planning Agency (BAPPENAS) 154, 157, 159; National Development Planning document (Yellow Book) 9–10, 154, 156, 157; National Medium-Term Development Plan (RPJMN) 155; nationally appropriate mitigation actions (NAMAs) 122, *124*; renewable energy *158*, *162*; sea levels 152; Second National Communication on Climate Change (SNC) 152, 161, 174n17; temperature increases 151–2; transportation *162*; water resources *158*, *162*; yen loan projects in power generation sector *200*

Indonesia Climate Change Sectoral Roadmap (ICCSR) 161

Indonesia–Japan Joint Statement for Enhanced Cooperation on Climate Change, Environment and Energy Issues 155

Indonesian Climate Change Trust Fund (ICCTF) 161

industry 249; China 105, 243; India 11; low carbon development policies *83*, *84*, 85

infrastructure projects 181–2

Institute for Global Environmental Strategies (IGES) 72

Integrated Gasification Combined Cycle (IGCC) plants 111

intercity collaboration 7, 14, 61–96, 250–1; bilateral 67–8, 75; and business development 78; financing 78–9; hub cities 70; incentives and motivations for participation in 74–5; and low carbon development 80–93; mutual referencing and learning 67–72, 81; platform secretariat 89–90; and sustainable development 75–80; technical cooperation 64, 65, *66*, *67*, 81–6, 90, *92*

Intergovernmental Panel on Climate Change (IPCC) 4, 20, 23, 28, 51, 99–100; Fourth Assessment Report (2007) 50; *Good Practice Guidance* 55, 56; guidelines (2006) 55

international city collaboration *see* intercity collaboration

international consultation and analysis (ICA) 24

international cooperation 116; *see also* intercity collaboration

International Council for Local Environmental Initiatives (ICLEI) 7, 62, 63, 64, *65*, *66*, 67, 85, 117, 250, 251

International Council for Local Environmental Initiatives (ICLEI) Europe 79

International Council for Local Environmental Initiatives (ICLEI) Global 80
International Council for Local Environmental Initiatives (ICLEI) Southeast Asia 68, *69*, 71, 73
International Development Association (IDA) 11–12, 210, 219
International Energy Agency (IEA) 115, 189
international framework 4–5, 19–27
International Institute for Sustainable Development (IISD) 75
International Local Government GHG Emissions Analysis Protocol 85
inter-regional collaboration 89
Iskandar Malaysia Low-Carbon Study 13, 227, 236, 237
Iskandar Regional Development Authority (IRDA) 237
Islamic Development Bank (IDB) 12, 210, 218

Jakarta 81
Japan 7, 27, 34, 59, 155, 243, 252–3; Action Plan for Achieving a Low-Carbon Society 12–13; City and Lifestyles Development Plan 234; coal-fired power **192**; earthquake (2011) 1–3; Eco Model Cities 85, 234–5; emissions reduction 12–13, *31*, 228, 229–31; energy efficiency 190, 192; Forestry and Forest Products Research Institute 59; local government engagement in technical cooperation *82*, 85, 86; low carbon society (LCS) development 81–9, 90–1, 227, 228–31; Mid- and Long-Term Roadmap Investigative Committee 230; Ministry of the Environment 228, 238; National Energy Development Organization (NEDO) 12, 222; National Institute for Environmental Studies (NIES) 12, 227, 237; natural gas **192**; nuclear power plants 1, 2, 15; official development assistance (ODA) 11, 15, 45, 75, 113, 181, 182, 193–201; oil-fired power **192**; Program Grant Aid for the Environment and Climate Change 59; Promotion Council for Low Carbon Cities (PCLCC) 93; solar power 221–2; Subsidiary Body for Scientific and Technological Advice 229
Japan External Trade Organization (JETRO) 86, 87
Japan International Cooperation Agency (JICA) 10, 11, 67, 86, 113, 155, 156, 157, 159, 160, 161, 165, 169, 237, 244, 253; technical cooperation programmes 81, *82*
"Japan Roadmaps towards LCSs" 228
Japan Science Technology Fund (JST) 237
Japanese Low-Carbon Society Scenario Research project 228–9, 231
JICA *see* Japan International Cooperation Agency (JICA)
joint carbon credit projects 81, 88, 92
Joint Implementation 180
Jordan 251

Kanagawa Prefecture *82*
Kathmandu *69*, 71, 72
Kawasaki 75, 77, 78, *82*, 85
kerosene consumption, Bangladesh 214–15, 216
Kitakyushu Asian Center for Low Carbon Society 86
Kitakyushu City 7, *69*, 70, 71, 72, 75, 78, *82*, 85, 86
Kitakyushu Initiative for a Clean Environment 7, 62, 63, 64–5, 67, 68, *69*, 71–2, 73, 74, 80, 250
Kitayushu Asian Center for Low Carbon Society 7, 93
Kiyono, Y. 56
knowledge networks 75–6
Korea 4, 19; Clean Development Mechanism (CDM) projects 43; emissions reduction target *31*, 122; nationally appropriate mitigation actions (NAMAs) 122, *124*
Kuala Lumpur *69*, 70
Kushiro City *82*
Kyoto City 12–13, 85, 227, 234–6, 238; Eco-model City Action Plan 235; Environmental Model City Action Plan 235; Global Warming Policy 234, 235; Research Team of Sustainable Society Kyoto 235
Kyoto Protocol 4, 22–3, 27, 29, 30, 37, 53–4, 164, 172, 180, 252
Kyoto University 12, 231, 232, 236, 237

Lake Biwa Environmental Research Institute (LBERI) 232, 233–4
land use 249
Large Cities Climate Leadership Group 251
livelihood improvement, electrification and 220

local governments 7, 14; Japan, engagement in technical cooperation *82*; low carbon development policies *83–4*, 231–6; *see also* intercity collaboration
local-level action 250–2; *see also* intercity collaboration; local governments
low carbon society (LCS) development 33, 61, 80–93, 227–40

mainstreaming climate change policies 150, 173n7
Makati 68, *69*, 70, *71*
Malaysia 11; Clean Development Mechanism (CDM) projects *43*; coal-fired power *183*, **191**, **192**; emissions reduction 196, *197–8*, 199; energy efficiency 195, 196; energy supply structure 189, 190; hydropower **191**; Iskandar City 13; natural gas 190, **191**, **192**; oil-fired power **191**, **192**, 195
Maldives, nationally appropriate mitigation actions (NAMAs) 122, *125*
Mandaue *9*
manufacturing industries, emissions **184**
Marrakesh Accords (2001) 22
Marshall Islands, nationally appropriate mitigation actions (NAMAs) 122, *125*
measurable, reportable, and veriable (MRV) systems 9, 13, 24, 26, 32, 58, 115, 120, 121, 126, 138, 142, 145
methane 108, 113
methane recovery 30, 43, *44*, 113
Mexico 19, 242; Clean Development Mechanism (CDM) projects 43, 246; nationally appropriate mitigation actions (NAMAs) 122, *125*; sectoral policy 249
Minamata *82*, 85
minimum energy performance standards (MEPS) 106
Ministerial Conference for Environment and Development in Asia and the Pacific (MCED) 64
mobile phone ownership, Bangladesh 212–13, 214, 215
MRV *see* measurable, reportable, and veriable (MRV) systems
Myanmar, deforestation *52*

Naga *69*, *71*
Nagoya City *82*
Nairobi Framework 47
nationally appropriate mitigation actions (NAMAs) 8–9, 13, 14, 23, 24, 26, 35, 115, 120–48, 242, 245, 247, 247–9; contribution of 122; financing 126, 142; methodology for developing: Thailand waste management case study 126–45; sharing results of 126
natural gas 189, 190, **191**; carbon intensity 185; China *191*, 192; India **191**, **192**; Malaysia 190, **191**, **192**; power generation 185, 188; thermal efficiency *190*, 194; Vietnam 190, **191**, **192**; yen loan projects *200*
Nepal 32
New Delhi metro project 112–13, 115
New York 251
Nigeria, deforestation *52*
nitrous oxide 108
nitrous oxide decomposition 5, 43, *44*, 45
non-Annex 1 countries 5, 24, 37, 241, 246, 247
non-governmental organizations (NGOs) 14, 15, 73, 117, 210–11
Nonthaburi *69*, 72, 78
Norway 58
nuclear power plants 185, 188; Japan 1, 2, 15

Obihiro City *82*
official development assistance (ODA) 11, 15, 45, 75, 113, 181, 182; coal-fired power plant projects 183–201; OECD 244
oil 29, 114, **189**; carbon intensity 185; China **191**, **192**; India **191**, **192**; Malaysia **191**, **192**, 195; thermal efficiency 190, 194; Vietnam **192**; yen loan projects *200*
Okayama University 236
ordered probit model 214, 215
Organization for Economic Cooperation and Development (OECD) 19, 182, 185, 203; Official Development Assistance 244
Osaka City *82*
Oslo Forest Climate Conference (2010) 58

Palli Bidyut Samity (PBS) 11, 205, 206, 209
Papua New Guinea 54, 59; deforestation 30; emissions reduction target *31*
Parry, Martin 149
Partners in Long-Term Urban Sustainability (PLUS) network 72
PDCA (plan, do, check, action) cycle 230
peat fire 152, *153*, **189**
Penang 68, 75, 78

per capita electricity production *186*
per capita emissions 3–4, 19, 28, 100, 185
Peru 247
Philippines: Clean Development Mechanism (CDM) projects *43*; yen loan projects in power generation sector 200
Phnom Penh 68, *71*, 78
Plan Vivo 246
policy-based approaches 172–3, 244–5, 248; *see also* climate change program loan (CCPL); development policy lending (DPL); sectoral policies
population: by region *186*; without electricity *188*
poverty reduction 100, 106, 107, 151
precipitation, Indonesia 163
PROCURA+ campaign 79
Program Grant Aid for the Environment and Climate Change 59
public–private partnerships: India 111; Thailand 142
Public Sector Accounting Reporting Standard 85
Puerto Princesa *69*, 71

quality of life 33

recycling 7; *see also* reduce, reuse, recycle (3Rs) initiatives
REDD+ Partnership 58, 59
REDD and REDD-plus 6, 13, 14, 23, 25, 30, 49–60, *158*, 161, 181, 245; methodology 54–7; policy related topics of 57–8; present state of 58–9; scientific background of 50–4
REDD Research and development Center 59
reduce, reuse, recycle (3Rs) initiatives 113, *129*, *139*, 145
reducing emissions from deforestation and forest degradation *see* REDD and REDD-plus
regional cooperation 32–4, 116
renewable energy: Indonesia *158*, *162*; subsidies 109, 110; *see also* biogas; biomass power; hydropower; nuclear power; solar power; waste-to-energy; wind power
renewable portfolio standards, India 109
reserve power 3
Ritsumeikan University 231, 232, 235, 236
roadmaps 232, 233, 239
Russia 27

Saitama City *82*
Saitama Prefecture *82*
Sapporo City *82*
science and policy interaction: local level 231–6; national level 228–31
Science and Technology Research on Environment Projects (SATREPS) 237
sea levels 27, 32, 152, 163
sectoral policies 99–119, 249–50
Seoul 68
Shanghai 250
Shenyang 75, 77, 78
Shiga Prefecture 12, 13, 227, 231–4, 238
simulation models 227, 230
Singapore 4; emissions reduction target *31*, 32, 122, *125*; nationally appropriate mitigation actions (NAMAs) 122, *125*
Singh, Manmohan 107
small and medium-sized enterprises (SMEs) 85, 86
smart energy grids 114–15
solar home systems, Bangladesh 11–12, 203–4, 209–24, 242, 247
solar panels: costs 221; manufacture of 31, 243
solar power 8, 29–30, 31, 34, 100, 114; China 8, 103; India 109; Japan 221–2; South America 247; *see also* solar home systems
South Africa 242; BASIC group membership 19; nationally appropriate mitigation actions (NAMAs) 122, *125*
South America 45, 51; *see also* Argentina; Bolivia; Brazil; Ecuador; Peru; Venezuela
static payback analysis 216–18
steel industry 105
Stern Review (2007) 51
structural adjustment lending (SAL) 150–1, 245
Subsidiary Body for Scientific and Technological Advice (SBSTA) 54–5, 58
subsidies, renewable energy 109, 110
Sudan, deforestation *52*
Surabaya *69*, 70, 71, 72, 78
sustainable development 5, 8, 22, 47, 122, 149–50, 173n6, 243–52; intercity collaboration and 75–80
Suzlon 109

Tanzania, deforestation *52*
technical cooperation 64, 65, *66*, *67*, 81–6, 90, 91, *92*

technology 27; appropriate 224, 247; performance standards 247; transfer 21, 23, 29, 30, 180, 224
Technology Mechanism 180
temperature, global 3, 14, 24, 26, 99, 149
Thailand 9, 237; Clean Development Mechanism (CDM) projects *43*; emissions/emissions reduction *139*, *140*, *141*, 145; nationally appropriate mitigation action (NAMA) 122; waste and wastewater management 127–47; yen loan projects in power generation sector *200*
Thailand Greenhouse Gas Management Organization (TGO) 138
thermal efficiency 189, 190–2, 194, 195–6
Tianjiin 250
Tokyo Electric Power Company (TEPCO) 1–2, 3
transportation 35, 115, 249, 250; China 106, 115; emissions **184**; India 112–13, 115; Indonesia *162*; low carbon development policies *84*
Tungsong 71
Turkey 242

Ulsan 72
UN-REDD 58
United Kingdom (UK) 23; Department for International Development (DFID) 165, 218
United Nations Development Programme (UNDP) 6, 58
United Nations Economic and Social Commission for Asia and the Pacific (UNESCAP) 72
United Nations Environment Programme (UNEP) 58; Eco Town project 78
United Nations Framework Convention on Climate Change (UNFCCC) 3, 19, 20–7, 29, 30, 31, 35, 37, 58, 107, 115, 152, 164, 172, 241
United States 22, 23, 27; and China Collaboration on Climate Change 251, 251; National Rural Electric Cooperative Association (NRECA) 204
United States Agency for International Development (USAID) 67, 204, 218, 219, 244
Universiti Technologie Malaysia (UTM) 13, 236, 237
urban environment management *82*
urban initiatives 116–17; *see also* intercity collaboration
urban planning 249, 250
Urban Sustainability Directors Network (USDN) 251

vehicles 115; compressed natural gas (CNG) 112, 245; emissions 112; fuel economy standards (FES) 106
Venezuela 25
Vietnam 9, 11, 122, 244; agricultural sector *167*; Clean Development Mechanism (CDM) projects *43*, 242, 244; coal-fired power *183*, **191**, **192**; development policy lending (DPL) to 163–71, 172; economic growth 164; emissions 164, *166–7*; emissions reduction 196, *197–8*, 199; energy consumption 164; energy efficiency 192, 195, 196; energy industry *166*; energy supply structure 164, 189, 190; forestation 51; hydropower 190, **191**, 199; industry *166*; land use and forestry 164, *167*; National Energy Efficiency Program 164; National Scheme on the Development of Biofuels 164; National Strategy for Natural Disaster Prevention, Response and Mitigation 164; National Target Program to Respond to Climate Change (NTP-RCC) 10, 165, 168, 171; natural gas 190, *191*, *192*; oil-fired power *192*; precipitation 163; sea levels 163; Support Program to Respond to Climate Change (SP-RCC) 10, 165–71, 172, 244; temperature increases 163
Voluntary Carbon Standards (VCS) 58–9
voluntary mitigation actions 241, 242–3; *see also* Clean Development Mechanism (CDM) projects; nationally appropriate mitigation actions (NAMAs)

waste management 7, 34, 35, *82*, 113, 250; Thailand nationally appropriate mitigation action (NAMA) 127–47
waste-to-energy projects: India 109, 113; Thailand *130*, *131*
water pollution *82*
water resources, Indonesia *158*, *162*
Weihai *69*, *72*
Willard, T. 76
wind power 43, *44*, 100, 114; China 8, 103, **104**; India 8, 109, 248; yen loan projects *200*
wind turbines, manufacture of 31
World Bank 6, 110, 150, 155, 161, 165,

World Bank *contd.*
169, 218, 244; Climate Investment Fund 115; Energy Sector Management Assistance Program (ESMAP) 110; forest Carbon Partnership Facility 58; International Development Association (IDA) 11–12, 210, 219; structural adjustment lending (SAL) 150–1, 245
World Resources Institute (WRI) 85, 115, 152, 164, 222

Yokohama 68, *69*, 70, 75, 78, *82*, 85
Yudhoyono, Susilo Bambang 155

Zimbabwe, deforestation *52*